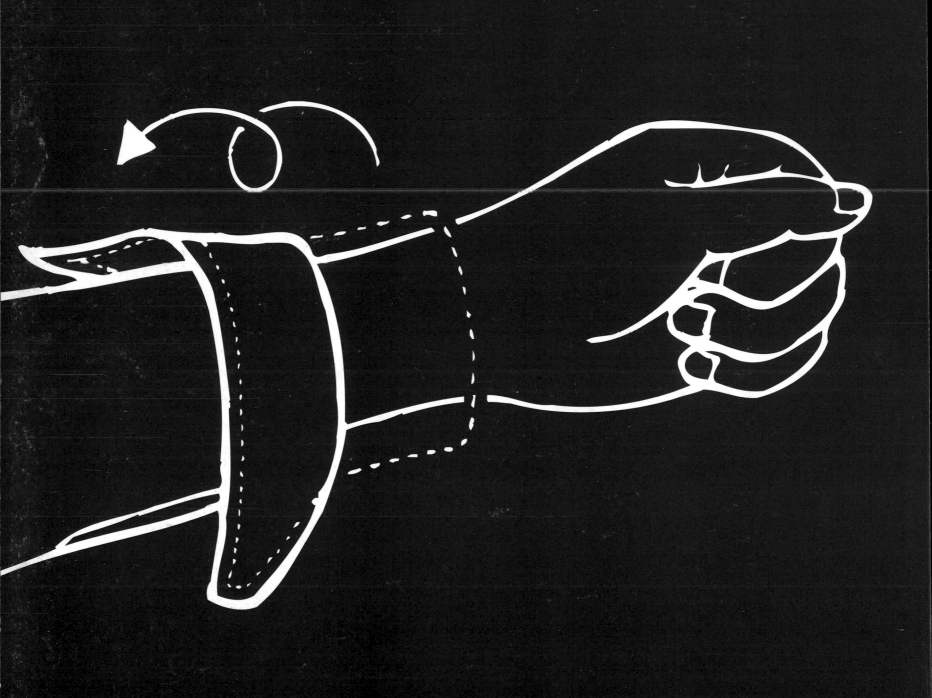

PAUL MIJKSENAAR   PIET WESTENDORP

OPEN HERE

THE ART OF INSTRUCTIONAL DESIGN

A JOOST ELFFERS BOOK

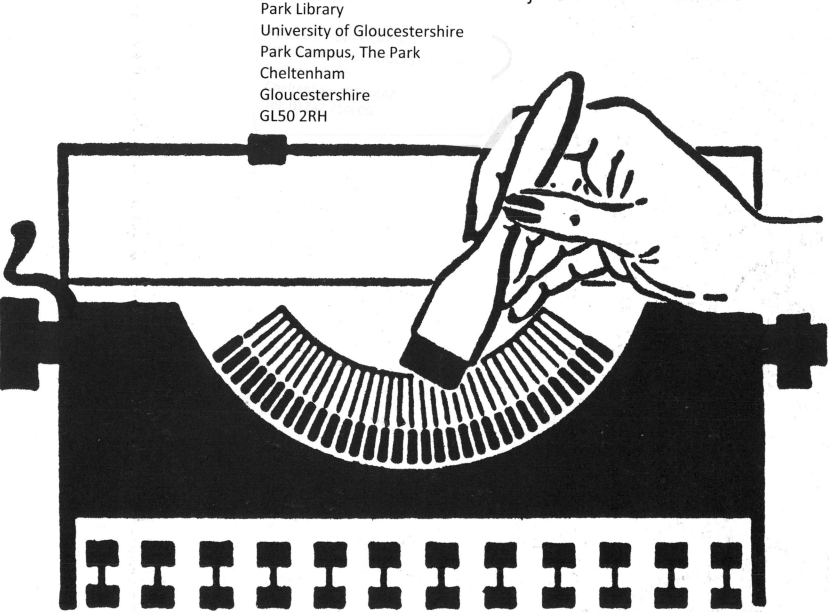

Thames & Hudson

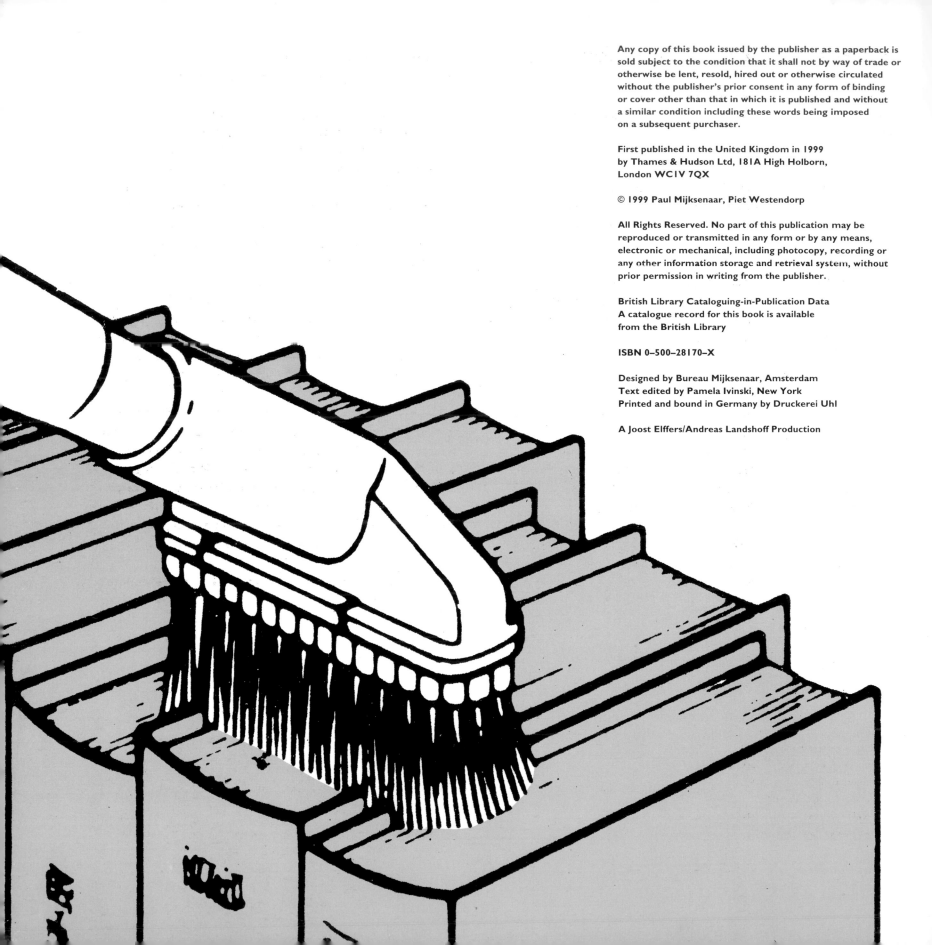

First published in the United Kingdom in 1999
by Thames & Hudson Ltd, 181A High Holborn,
London WC1V 7QX

British Library Cataloguing-in-Publication Data
A catalogue record for this book is available
from the British Library

ISBN 0–500–28170–X

Designed by Bureau Mijksenaar, Amsterdam
Text edited by Pamela Ivinski, New York
Printed and bound in Germany by Druckerei Uhl

A Joost Elffers/Andreas Landshoff Production

Please read carefully before . . .

On a traffic sign near an intersection in Chicago, the complications of modern life are summarized concisely. The sign says: "To turn left, turn right twice."
Edwin Way Teale

Congratulations on your purchase of *Open Here: The Art of Instructional Design*. We're sure you will appreciate it. But before using this book, please make sure you have studied its contents carefully.

Open Here has been designed and manufactured to the highest standards. It will show you around your daily Kafkaesque life of incomprehensible technology, frustrating packaging, do-it-yourself disasters—and the Visual Instructive Esperanto that should help you out.

Open Here is for all those who appreciate everyday art with a lowercase "a." It is also meant as a source of inspiration for those who produce such art, to help them find smart—or shrewd—solutions to the communication problems they face when they have to tell us where to look, how to twist and what *not* to do. *Open Here* presents designers and illustrators with the ideas and clever tricks of their professional predecessors and colleagues.

This book reveals how much we depend upon visual instructions in daily life. We consult maps, school books, traffic signs, training manuals and scientific illustrations. In *Open Here*, we focus on the visual instructions that help us to solve the most basic problems of each day: how to open a child-proof bottle of aspirin, make a reduced-size double-sided photocopy, listen to voice mail using a mobile phone, program the microwave oven to turn on automatically and have dinner ready when we arrive home. This book contends that products definitely do not speak for themselves and that it's only getting worse. It also surveys a wide range of solutions that designers and illustrators can employ in creating visual instructions: mediums they can use, concepts they can apply and various ways they can communicate to different target groups. The book then looks at inventive instructional elements in detail, more or less in the order they should be produced and used, from warnings and identification of parts to composition and connections through the movements and sequences that must be performed and finally, the successful results: a VCR that works!

We feature the lucky strikes of anonymous artists. Ingenious solutions to tricky communication problems. We do not judge their effectiveness or beauty, but merely show what surprised us as original, creative, thrilling, touching or simply unbelievable. You may find examples from a manual that drove you crazy.

We collected these illustrations from manuals, cut them from packages, copied them from museum and company archives, and stole them from airplanes. Too often, we never even saw the products that accompanied the illustrations. We've enlarged some pieces and added or changed colors for a few others. At certain points, the images have been arranged to facilitate comparison. But in general we give you these illustrations without prettifying or perfecting them—the authentic pictures are alienating enough.

The language of visual instructions remains very primitive, with a limited number of signs and a weakly developed grammar. Visual instructions are still a small sub-category of visual information. A dialect of simple utterances: *Attention! Press. Turn. Push. This side up. Take care. Open here!*

Note: read carefully all the pages of this booklet, especially the chapter "Read carefully before using Cellesse."
From the Philips Cellesse cellulite massage system manual

1. Sit down. 2. Light cigar at correct end (the pointed one). 3. Open book. 4. Enjoy.

Thanks . . . !

We would like to thank all the people, companies, museums (and airlines) that helped us to build the giant collection of visual instructions on which this book is based. We would also like to thank all the companies that hire designers, art directors and illustrators to reinvent the communication wheel over and over again. Our world of technology would be naked and impassable without them.

Many, many friends and relatives, colleagues and business partners have put their manuals at our disposal, sometimes even voluntarily. Thousands and thousands we have. But we're still greedy. So please continue sending your beauties to us at: Delft University of Technology, Faculty of Industrial Design Engineering, P.O. Box 5, 2600 AA Delft, the Netherlands. You don't need them anyway—you think.

We have done the best we can to locate the rightful claimants to the illustrations in this book. But sometimes they were too anonymous to find. Those who believe they have rights to these illustrations are invited to contact the Archive Paul Mijksenaar Foundation, Haaksbergweg 15, 1101 BP, Amsterdam, the Netherlands.

Most important, this book is a tribute to the "Unfamous Artists" who produced these little pieces of great day-to-day instructional art. The Roy Lichtensteins and Andy Warhols without name—the people who usually get blamed when we can't get that VCR to record what we want, even though the problem started long before they were involved.

● Fanny de Boer; Annemarie van den Bos; Henk van Dijk; Nadya Glawé; Martine Houwert; Elise de Jong; Julia van Leeuwen; Debby Lindeboom; Radha Pancham; Linda Pilgrim; Elvire Schukken; Tammy Stubbs; Ton Veldkamp; Students of Delft University of Technology

● DAF Museum, *Eindhoven*; Scryption, Museum voor Techniek en Vormgeving van Schrift en Kantoor, *Tilburg*; Rail-Toy Speelgoedmuseum, *Oostvoorne*; Speelgoed- en Blikmuseum, *Deventer*; Naaimachinemuseum, *Dordrecht*; Speelgoedmuseum Op Stelten, *Oosterhout*; Legermuseum, *Delft*; Philips Company Archives, *Eindhoven*; Mercedes-Benz Company Archives & Library, *Stuttgart*; Delft University of Technology, *Delft*; Archive Paul Mijksenaar Foundation, *Amsterdam*

● 3M; AEG-Eurolec; Aeroglisseur; Agio sigaren; Ahrend; Air France; Air UK; Akai; Akco; Akzo; Albert Heijn; Allergan Pharmaceuticals; Pat Andrea; Anker Steinbau; ANWB; Apple; Arlac; Aronal; Aspa; B+H; BC Ferries; Beecham Farma; Beecham Research Laboratories; Bemico; Bialetti; Bilosa; Boudewijn Bjelke; Black & Decker; Bonduelle; Bosch; Braun; British Airways; British Caledonian; British Midland; British Telecom; Brother; C&A; Caledonian Airways; Cambo; Canon; Chefaro International; Citroën; Coleman; Coventry & District Co-op Society; Crossair; Daimon; Dan Air; Darda; De Bijenkorf; Dorma; Douwe Egberts; Durex; Dylon International; Eduardo de Lima Castro netto; Elmex; Erkende Verhuizers; Escupac; Esselte Letraset; Euro-Grill; Evian; Excell Enterprises; F.G. van den Heuvel; Faber-Castell; FGB; Fiat; Filtane; Finnair; Fisons Pharmaceuticals; Flexible Lite; Ford; Fuji; Gaba; Garcia; General Motors; Gesellschafts- und Wirtschaftsmuseum, *Wien*; Gestetner; Gillette; Glaxo; Goodman; Grundig; Guernsey Airlines; Haag Techno; Hello Yello; Hema; Hewlett Packard; HMSO; Holiday Inn; Hotel Janpath; Hoverspeed; Hunkler; Husqvarna; IBM; Ikea; Interflug; Intergamma; International Student Travel Conference; Interstuhl; JAL; Janssen Pharmaceutica; JVC; Kalorik; Kaptein; Kibri; Kleine; KLM; Kodak; KPN Telecom; Krups; Landelijk Contactorgaan Begeleiding Borstkankerpatiënten; LEGO; LGB; Lipton; Lufthansa; Lundia; McDonald's; Macintosh; Maersk Air; Magnavox; Malaysia Airlines; Malèv Hungarian Airlines; Marantz; Märklin; Matador; Mattel; Mecano; Medico-pharma; Merck Sharp & Dohme; Merpati; Micki; Migros; Moha; Mols-Linien; Mullard Limited; MW; National; Natterman; Nestle; Nikon; Nilfisk; Nissin; Non Stop Box; Novartis Consumer Health; NS; NTT; NWT Air; Océ; Olfa; Olympus; L'Oréal; Panasonic; Parker; Petzl International; Philips Electronics; Pickwick; Plastmeccanica; Polaroid; Poly Color; Power Prop; Procter & Gamble; PTT Museum, *the Hague*; Pyrex; RATP; Realistic; Reckitt; Revell; Roberto; Rombouts; Rotpunkt; Rotring; Roussel; Royal Air Maroc; Royal Brunei; Royal Jordanian; Royco; S.A. Les Editions de Saxe; Saba; Sanyo; SAS; Seidel & Naumann; Shell; Siemens; Sigma Sport; Silk Air; Singapore Airlines; Singer; Skil; Smith & Nephew Pharmaceuticals; Soframycine; Sony; Staatsuitgeverij; Stanley; Stichting Vi-taal; Storchenmühle; Sturmey Archer; Suzuki; Swatch; Swissair; Tailored Neckwear; Tajima Industry; Tamiya; TDK; Technische Unie; Telemalta Corporation; Tie Rack; Tip Top; Transamerica Airlines; Transavia Airlines; United Airlines; US Air; VFW-Fokker; Volvo; VW; Warrick; Wasa; Wenger; Westmark; Winthrop; Wooster; Xerox; Yamaha; Zasex; Zyma . . . and many many others.

Overview

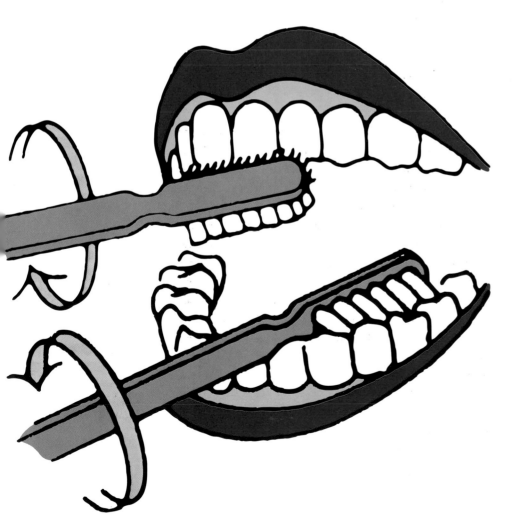

Modern life is a continuous intelligence test.

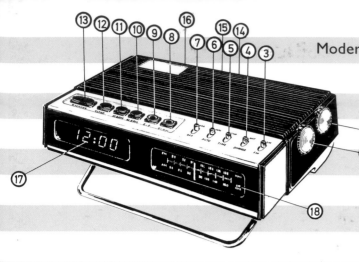

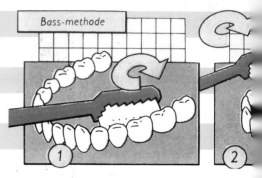

snooze? Time to brush our teeth

shave (*To clean the hairs from the*

simultaneously back and

radio (*Press pre-set button 1*) and the air conditioning (*For extra cooling,*

select level 3). Heed

box of paper clips—oops, on the

floor because

press. . . . Switch on the computer by

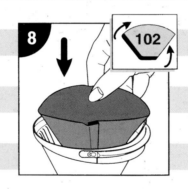

to start the coffee maker? Of course this toaster

teeth (*Move the thread up and*

down. . . .). Back to the calculator, the word

your telephone with a memory for 200 numbers, a voice response system, automatic call back (*Press *. . . .*) and dozens of

vending machine and find the

platform. First a telephone

ticket). **Home,** sweet

home. With a microwave

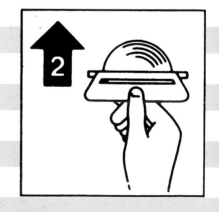

liquid

detergent. . . .). But

memory scan function. . . .) and program the VCR (*For channel, press. . . .*).

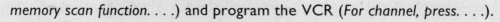

The real wizards will now switch on their computers and log on to the Internet. **Time** to get

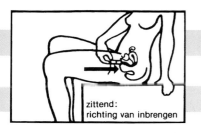

zittend:
richting van inbrengen

clock radio till the moment we set it again for the next morning. Which button for

movements). Insert a tampon (*See picture*) or

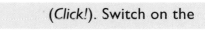

PURE-PAK®

Hier   drücken

knot in your tie, follow the instructions. To open the milk carton, press the two sides

he car. Start the engine (*Turn the key to position 1*), put on the seat belt

(*Click!*). Switch on the

et in: Insert your card with the arrow up and

forward. Open a new

Finding a fax number in your new

digital diary is easy; first

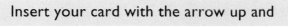

Time for a lunch break (*Press*

F10 to save). But how to

the plastic

cheese wrapper, the jam jar? How

from the one at

home. Floss your

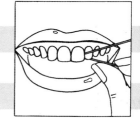

the "Busy" light is

on, press. . . .) and

here), so take public transportation this

time. Buy a subway ticket from the

HIER OPENTREKKEN   TENMINSTE HOUDBAAR TOT:
00.05.94

jong belegen 48+

Kaptein - IJmuiden - Holland

station (*1. Choose your destination. 2. Insert*

money—or bank card? 3. Remove your

mash potatoes, hold

down button 1 and press. . . .), the washing machine (*Pour in*

stereo (*To select a*

station with the

the cupboard assembly

kit together.

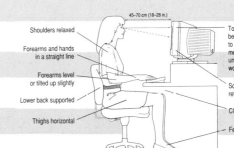

45–70 cm (18–28 in.)

Shoulders relaxed

Top of the screen at or slightly
below eye level (You may need
to adjust the height of your
monitor by placing something
under it or by raising your
work surface.)

Forearms and hands
in a straight line

Forearms level
or tilted up slightly

Lower back supported

Screen positioned to avoid
reflected glare

Clearance under work surface

Thighs horizontal

Feet flat on the floor

m (*Take the condom between thumb and forefinger and roll it gently. . . .*).

The Daily Intelligence Test

There are two kinds of truths. Those of reasoning and those of fact.
Gottfried Wilhelm von Leibniz

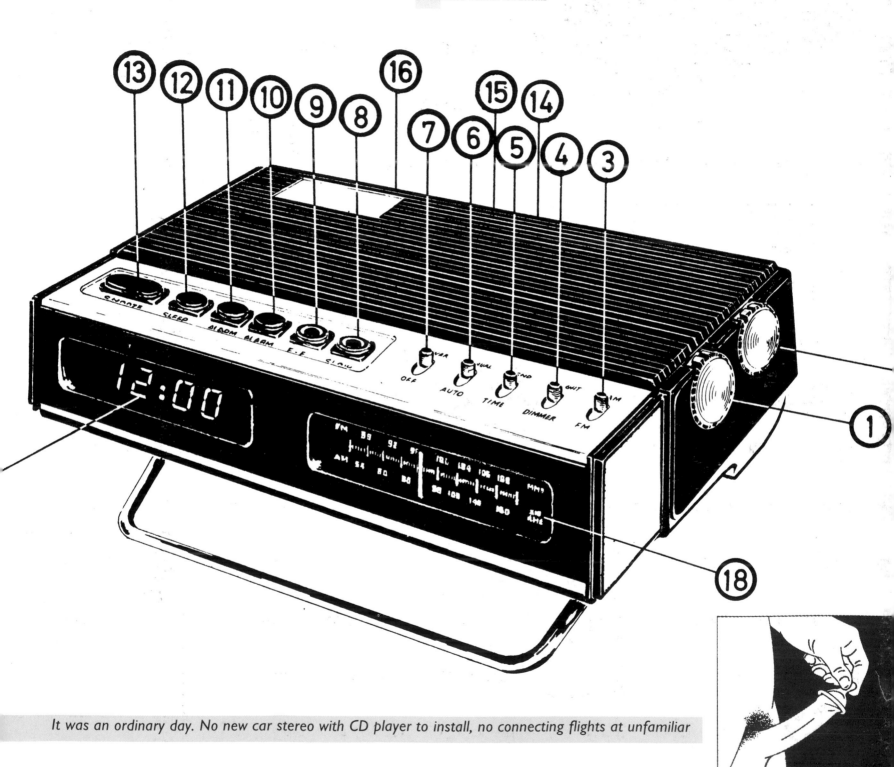

It was an ordinary day. No new car stereo with CD player to install, no connecting flights at unfamiliar

airports, no really intricate packaging. It was an easy day. No need to use graphic language.

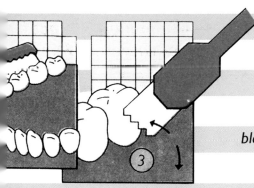

From the moment we  **wake up** and switch off the

(. . . horizontal, vertical and *turning*

blade, push the tabs. . . .). To make a

pull. . . . Set the toaster on light brown instead of burned black (*Turn knob A to position 3*). **Into t**

traffic lights, road signs, traffic police and pay the parking meter (*First insert coins, then turn the knob*). **At the office,** but not y

we missed the up-arrow. Make five photocopies, double-sided and sorted. But of course the light is on.

pressing. . . . Put through a telephone call (*First press button A, then dial the internal number*).

open the cup of oriental noodle soup,

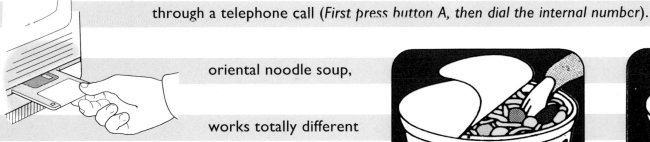

works totally different

processor (*Press F5 to see the directory*), the printer (*If*

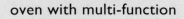

other functions. **Time to leave.** Car to the service garage (*Enter*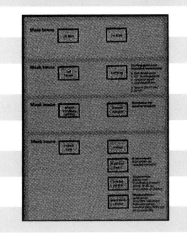

call (*insert the card—but where and how?*). Next challenge: the ticket machine at the railway

oven with multi-function timer, the food processor (*To*

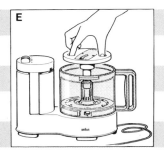

how to open the bottle with the child-safe lid? Turn on the

Congratulations on your new dishwasher. After managing to open a pack of diapers, start putting

ready for bed. Use the Water Pik, and some nose drops (*See insert*), set the clock radio again to 7:00 AM and maybe use a condo

"I just paid $2,000 for this damn thing, and I'm not going to read a book."

The exasperated help-line caller said she couldn't get her new Dell computer to turn on. Jay Alblinger, a Dell Computer Corp. technician, made sure the computer was plugged in and then asked the woman what happened when she pushed the power button.

"I've pushed and pushed on this foot pedal and nothing happens," the woman replied. "Foot pedal?" the technician asked. "Yes," the woman said, "this little white foot pedal with the on switch." The "foot pedal," it turned out, was the computer's mouse, a hand-operated device that helps to control the computer's operations. . . .

. . . a customer was having trouble reading word-processing files from his old diskettes. After troubleshooting for magnets and heat failed to diagnose the problem, [the technician] asked what else was being done with the diskette. The customer's response: "I put a label on the diskette, roll it into the typewriter. . . ."

. . . At AST, another customer dutifully complied with a technician's request that she send in a copy of a defective floppy disk. A letter from the customer arrived a few days later, along with a Xerox copy of the floppy.

. . . And at Dell, a technician advised a customer to put his troubled floppy back in the drive and "close the door." Asking the technician to "hold on," the customer put the phone down and was heard walking over to shut the door of his room. The technician meant the door to his floppy drive.

. . . A Dell customer called to say he couldn't get his computer to fax anything. After 40 minutes of troubleshooting, the technician discovered the man was trying to fax a piece of paper by holding it in front of the monitor screen and hitting the "send" key.

. . . A Dell technician . . . says he once calmed a man who became enraged because "his computer had told him he was bad and an invalid." [The technician] patiently explained that the computer's "bad command" and "invalid" responses shouldn't be taken personally.

From Jim Carlton, "Befuddled PC Users Flood Help Lines, And No Question Seems to Be Too Basic," *The Wall Street Journal,* March 1, 1994, pp. B1, B6.

And then there was the man who complained that the tray for his soft drink cans had broken off again. Meaning the CD-ROM tray. . . .

You Can't Go Wrong

Before the Law stands a doorkeeper. A man from the country comes to this doorkeeper and requests admittance to the Law. But the doorkeeper says that he can't grant him admittance now. The man thinks it over and then asks if he'll be allowed to enter later. "It's possible," says the doorkeeper, "but not now."

Franz Kafka, *The Trial* (1925)

Before the VCR there is a manual. . . . Franz Kafka actually wrote instructions for the operation of dangerous industrial machinery in order to prevent accidents and thus reduce the number of claims made to the insurance company where he worked.

"Take the first right, go straight at the traffic circle, and turn left at the second street after you cross the railroad tracks. You can't miss it." You can't? The man who just gave you directions obviously doesn't realize how many ways there are to go wrong. "Should I count this alley as a road?" "Did he say left or right?" "Before or after the railroad tracks?" In fact, you can miss it in an incredible number of ways. If you didn't, if you made the right choice at every point, it was probably sheer luck.

The same is true for high-tech electronic products. You just bought a mobile phone and the salesperson assured you: "Really, it speaks for itself." Let's hope not—if it could speak, it would more likely pick a fight than offer suggestions on how it should be used. Even if you only have to struggle with their buttons, icons, display messages and manuals, most high-tech products are not the least bit self-explanatory. This is not only because their interfaces and instructions for use may be badly designed. It's simply impossible for a product that offers more than 100 features to be completely self-explanatory without an instructional manual. And once you have to follow instructions, you *can* go wrong.

Products never speak for themselves. Someone had to teach us that a chair is meant for sitting on, that a spoon is for putting food into the mouth. How to use these things seems intuitive because we learned about them at an early age. But imagine seeing a flat piece of curved wood with rounded corners for the first time. Would you guess what a boomerang is for? And how to use it? Would somebody who had never seen a fork before understand that it is meant for use in eating?

John looks at the motorcycle and he sees steel in various shapes and has negative feelings about these steel shapes and turns off the whole thing. I look at the shapes of the steel now and I see *ideas*. He thinks I'm working on *parts*. I'm working on *concepts*.

Robert M. Pirsig, *Zen and the Art of Motorcycle Maintenance* (1974)

Often we cannot guess the function of even the simplest products of our own cultural environment. Let alone figure out how to use them.

Look at the simple, non-electronic kitchen devices on these pages. Do you know what they are? And how to use them? It should be easy now that you know they're all kitchen tools. Imagine how difficult this task would be if these products could also be useful gadgets for clothing repair or fishing. Of course, some of them might prove quite handy for that. . . .

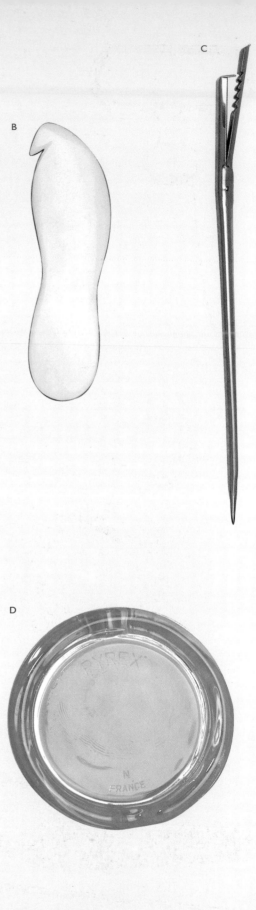

B

C

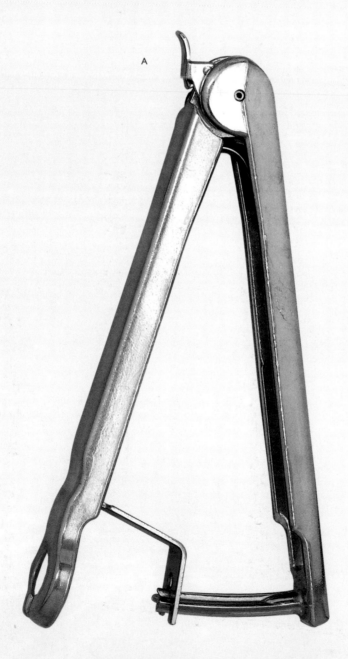

A

D

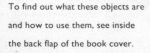

To find out what these objects are
and how to use them, see inside
the back flap of the book cover.

The Art of Instructional Design

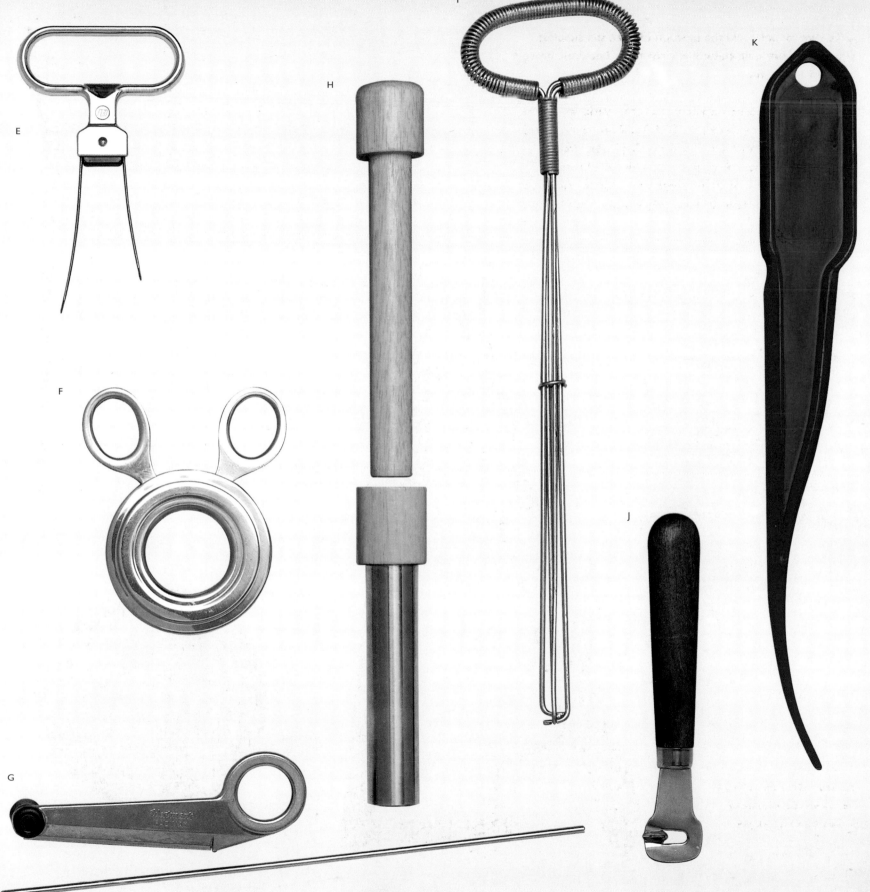

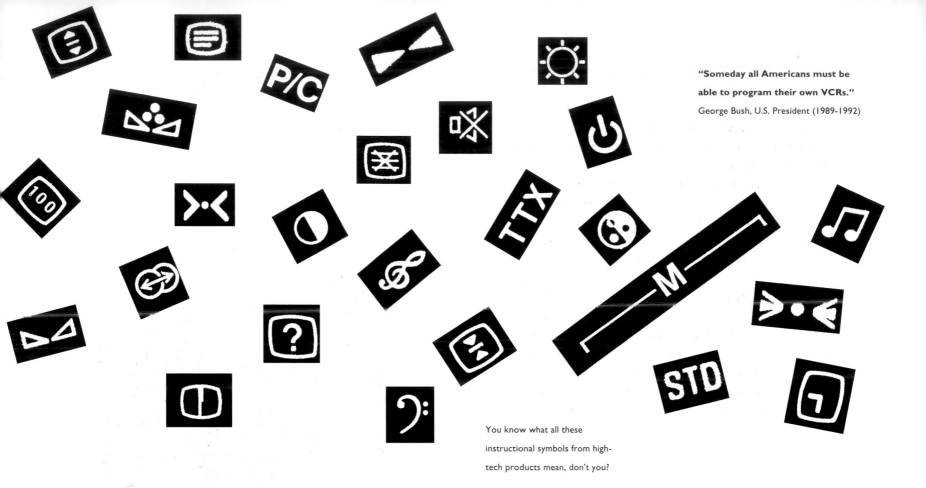

"Someday all Americans must be able to program their own VCRs."
George Bush, U.S. President (1989-1992)

You know what all these instructional symbols from high-tech products mean, don't you?

So much for simple products. How about playing the same little game with a mobile phone-fax-e-mail unit, a world-band radio or an electronic organizer? Could you guess what they are? Understand how to use them, including any hidden functions? Do they speak for themselves? No way. Only to the specialists who sell them—*they* know what you can do with these high-tech products, and what you shouldn't. They're familiar with all the instructions and icons. But you're not, so you get a manual—180 pages of technical gobbledygook, including exploded-view illustrations with gymnastic arrows, funny magnifying glasses, 82 reference numbers and childish anthropomorphic cartoons to warn you about the most incredible situations. You're on your own now.

Will having to rely upon these complex manuals stop us from buying products with more and more features? No, we'll go on buying them. And we'll go on not using them. A study once showed that 95% of Ricoh fax machine owners did not use the machine's three most important extra features. But we want them anyway. Because they're cheap, because they seemed handy when demonstrated, because we wanted *one* of them (and they always come in groups), because the neighbors have them.

The latest development is to make high-tech products look easy again. The newest remote controls, cameras and radios have very few buttons. But the problems are not solved, just hidden. Sometimes the trick is too obvious, as with remote controls and copiers that seem to have very few buttons—till you look at the back, move a lid or open two small doors to discover that the product is loaded with buttons. Even worse are the machines with screens. Now you really have a limited number of real buttons because all the features and settings are hidden in menus, commands, soft keys, screen buttons and icons. It *looks* easier, but it's far more complicated. Now we need wizards, guides and on-line help systems to find our virtual way in feature-land.

"You can put the warning in neon light, but they still won't read it."

Mixers are sold with two beaters and two dough hooks. The beaters are exactly the same, but the left and right dough hooks are different. The left one must be placed in the left opening and the right one in the right opening of the mixer, or else the machine will swing dough all over the kitchen (try Speed "III"). So there was a **WARNING** in the manual: "Be sure to put the left dough hook in the left opening," etc. That didn't help, even when pictures were included. So the company put a small red ring around the left dough hook and a red circle around the left opening. This didn't help, either. People still complained that this horrible machine kept flinging dough around their kitchens. So the company decided to change the design of the product. The left opening was made a bit wider and the left dough hook was made accordingly thicker. Now it was impossible to put the dough hooks in the wrong openings. Problem solved, right? Well. . . . Now some people asked why they had to put one specific *beater* in the left opening and the other in the right opening. Wouldn't it be much easier if each beater could be placed in either opening? The beaters were exactly the same, were they not?

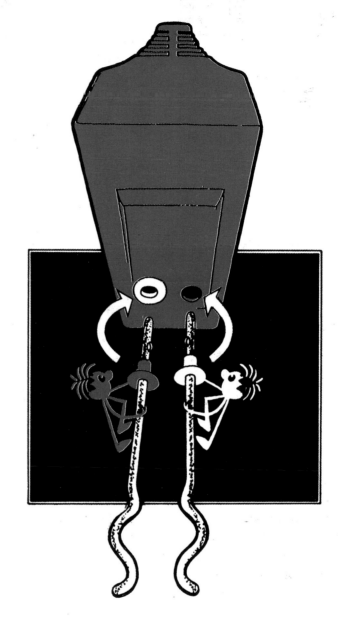

"One fool can ask more questions than a thousand wise men can answer."
And a telephone with 200 features is sold to millions of fools.

High-tech Toilets

An American diplomat was at a dinner party in a Japanese home when he excused himself to go to the bathroom. He did his business, stood up and realized he didn't have a clue about how to flush the toilet.

The diplomat speaks Japanese, but he was still baffled by the colorful array of buttons on the complicated keypad on the toilet. So he just started pushing.

He hit the noisemaker button that makes a flushing sound to mask any noise you might be making in the john. He hit the button that starts the blow-dryer for your bottom. Then he hit the bidet button and watched helplessly as a little plastic arm, sort of a squirt gun shaped like a toothbrush, appeared from the back of the bowl and began shooting a stream of warm water across the room and onto the mirror.

And that's how one of America's promising young Foreign Service officers ended up frantically wiping down a Japanese bathroom with a wad of toilet paper.

From "But Do They Flush?" by Mary Jordan and Kevin Sullivan, *The Washington Post*, May 15, 1997, p. A1.

Treatises to Technofrenzy

There is no reason for any individual to have a computer in their home.

Ken Olson, president, chairman and founder of Digital Equipment Corporation, 1977

The illustrations in Hans Thalhoffer's *Fechtbuch* [Fight Book] (1443) instruct readers on the arts of unarmed combat.

Visual instructions are by no means a recent invention. Since the Middle Ages, books and treatises have illustrated how to perform some of the tasks of daily life. First, let's look at today's two audiences for user instructions, technofreaks and technophobes, and then briefly trace the evolution of visual instructions.

Everything that can be invented, has been invented.

Charles H. Duell, Commissioner, U.S. Office of Patents, 1899

Technofreaks and Technophobes

British physicist and writer C. P. Snow may have been the first to notice the growing gap between scientists and intellectuals. In his 1959 lecture "The Two Cultures" he criticized the lack of understanding and the profound mutual suspicion between scientists and intellectuals, which diminish the possibilities for using technology to solve humanity's most serious problems.

Since then, the gap has widened. It's no longer just between scientists and intellectuals, but also between two new cultures: technofreaks and technophobes. Technofreaks buy a brand-new motorcycle and immediately take it completely apart. They love phone-fax-answering machines with dozens of functions—and can't wait to buy a newer version. Naturally, technofreaks always have the latest word processor installed—so they can include sound in their electronic letters and make hyperlinks to their websites. They have electronic organizers with calculators and dictionaries and they buy digital video cameras just to try out the latest functions. Making

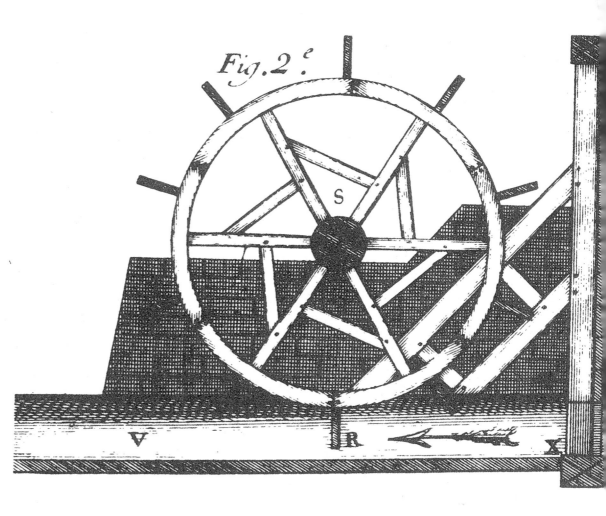

During the Middle Ages, labels and captions
were inserted extensively in instructional
pictures to explain elements. Leonardo da Vinci
was the first to insert letters in his scientific
drawings to which he referred in his texts. He
derived this method from geometrical
demonstrations in the tradition of Euclid. His
reference system was soon generally adopted.
Based on Ernst Gombrich, "Pictorial Instructions," in
Images and Understanding, Barlow, Blakemore and
Weston-Smith, eds., 1990.

Short History of the Metaphoric Arrow

Today arrows are widely used as indicators of
direction and movement. But where and when did
the arrow get its universal symbolic significance as
a pointer or vector?

The first symbolic use of the arrow appears in the
compass rose, introduced by the ancient Greeks
circa 150 B.C. Astronomers of that period realized
the earth is a sphere turning around the sun, and it
was probably Hipparchus who suggested
representing the planet with an imaginary network
of parallels and meridians (latitude and longitude).
Hipparchus may also have been the first to use
arrowheads on the compass rose, which is said to
represent the sun with an arrow indicating north.
By way of the compass rose and compass (the
earliest European mention of which dates to the
12th century), the arrow was introduced into
cartography. In the 18th century we see the first
arrows used on maps, to indicate, for instance, the
direction in which a river flows. Arrows first appear
in technical drawings early in the 18th century, the
earliest example being an arrow used to indicate
direction in a waterwheel diagram from *Architecture
hydraulique* by Bernard Forest de Bélidor, 1737. In
the latter half of the 19th century, arrows were
sometimes drawn on sewing machine wheels to
indicate the direction in which they should be
turned. By the beginning of the 20th century arrows
were already being used in traffic signs. The more
realistic hand with a pointing finger, which had been
employed since the 17th century to indicate
direction, had definitively lost out to the arrow.

The shape of symbolic arrows has evolved from
realistic to stylized. At first, arrows were barbed
and had feathered shafts, just like the arrows used
for hunting. Later, especially in the 20th century,
they have become stylized, losing all realistic
elements. It is only during the last few decades that
we see three-dimensional and twisting arrows, used
to indicate complex movements. Moreover, arrows
now fulfill metaphorical functions: they not only
indicate direction or location, but also movements,
size, distance and order (e.g. in a flowchart).
Based on Otl Aichler and Martin Krampen, *Zeichensystemen
der visuellen Kommunikation*, 1977.

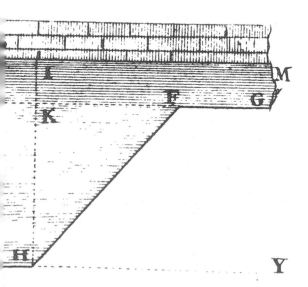

home movies doesn't interest them much. The features, the functions, the possibilities, the power—that's what counts.

Technophobes love writing letters, meeting people on time and making home movies. They hate buttons, displays, on-line help and manuals. Despite struggling through a product's 168-page manual ("An experience designed to humble and humiliate people") and despite trying their best to decipher its funny little pictures, they still press the wrong buttons. And they always manage to press the only one with the power to violate the machine.

The technophobe's nightmare may have started with the Industrial Revolution, but it's only since the Information Revolution that instructions have become so indispensable. Before the Industrial Revolution, a knife was just a knife, not a Swiss Army knife with 34 attachments and 85 functions. A bottle had a cork, not a plastic child-safe push-and-turn cap. And writing was done with a pen, not with Microsoft Word version 9.01 on a computer with a modem and Internet connection. We didn't need instructions to tell us how to make tea, how to inflate the soles of our shoes or unpack a washing machine, what to do in case of emergency on a plane, how to unwrap a slice of cheese or insert a bank card into an ATM. Before the Information Revolution, we did not have to be engineers to play with our children's toys.

When Life Was Easy

"What writing is to the reader, pictures are to those who cannot read," stated Pope Gregory the Great (540-604), which reminds us that visual arts in the Christian Church have been assigned the task of pictorial instruction, to avoid charges of idolatry. By the Middle Ages, a wide spectrum of image types—from realistic drawings to purely abstract diagrams—was already being used for teaching and training. Examples range from an illustration in a medical treatise that shows the points of the body suitable for cauterization to images depicting the forms for holding a bird on the fist in a treatise on the art of hawking. Pictures were used for the identification of herbs, for instruction on how to load and use a cannon, how to swim or fight with a sword, and for training in horseback riding and angling. Visual instructions

Singer central bobbin sewing machine. United States, around 1920.

The Art of Instructional Design

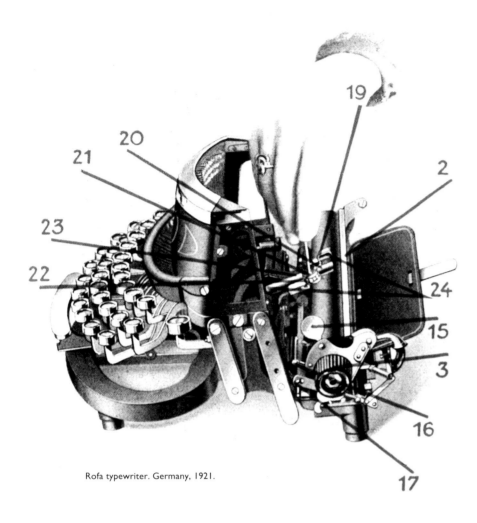

were also employed to guide the builders of ships, churches, windmills and battering rams. In fact, there is hardly an aspect of life in those times on which a manuscript treatise does not survive.

Manuals have been preserved since the 15th century. Many are strictly utilitarian, lacking any artistic ambition, such as *Fechtbuch* [Fight Book], a fencing manual by Hans Thalhoffer (1443), which includes instructions in wrestling and unarmed combat. A very sophisticated handbook, *Wapen handelinghe van roers, musquetten ende spiesen* [How to Use Firelocks, Muskets and Spears] by Jacob de Gheyn, was first published in 1597. Another fine example, *Orchésographie*, by Thoinot Arbeau (1589), uses the motions of fencing to explain the movements of dancing. M. Giegher's book on the art of catering, *Trattato delle Piegature* (1639), includes instructions on napkin folding, which may be some of the first to illustrate the exact positions of the hands in performing the task in addition to the desired result.

Many of the rhetorical features of these books resemble the practices of modern technical writing. For instance, the *Treatise of Fishing with an Angle* (1496) makes an effort to relieve user stress about the new technology it introduces and employs images to supplement the text in ways that seem quite modern. The treatise includes seven illustrations (and refers to an eighth, which was never printed—in this way, too, the treatise is very similar to modern instructions for use!). In addition, references to these illustrations are made in the text, images are placed near the appropriate technical discussion and call-outs are used as labels.

Probably the greatest early enterprise in pictorial instruction is *La Grande Encyclopédie*, launched in 1751 by Diderot and D'Alembert, who planned to publish 600 pages of instructive illustrations in order to remove secrecy from craft traditions.

Rofa typewriter. Germany, 1921.

Sirs, I have tested your machine. It adds new terror to life and makes death a long felt want.
Sir Herbert Robert Beerbohm Tree on examining a gramophone

This text is based upon Ernst Gombrich, "Pictorial Instructions," in H. Barlow, C. Blakemore and M. Weston-Smith, eds., *Images and Understanding*, Cambridge University Press, 1990, pp. 26-45; and Max Loges, "The Treatise of Fishing with an Angle," *Journal of Technical Writing and Communication*, 24/1 (1994), pp. 37-48.

The Age of the Engineer

And then the Industrial Revolution arrived, with its mass-produced sewing machines, typewriters, telephones, radios and cars. Suddenly people had machines they could operate only by consulting instructions. These were the first products with manuals as we know them: instructions for a specific product (for instance, the Singer Type I sewing machine of 1851) instead of a category of products (sewing machines in general). A person who bought a sewing machine or a typewriter would get instructions from the manufacturer or reseller, often in a classroom setting. Manuals provided general information and repetition of lessons. And, of course, they were small compared to our current magnum-size books. The complete documentation of the Argus Flugmotor (a German airplane engine) of 1916 was 85 pages; the total technical documentation of the Boeing 747 weighs more than a Boeing 747.

In 19th-century manuals, illustrations were primarily realistic; some were abstract to convey general technical principles. In fact, the visuals in 19th-century manuals were basically of the same type as those introduced by Leonardo da Vinci in the 15th century: an overview picture of the complete product with numbers corresponding to details mentioned in the text. Very rarely do we see illustrations of product details, and the only instructional elements used were hands shown holding parts of the machine and some simple arrows. From an instructional point of view, there was not much development in visual instructional language from the 15th until the 20th century.

The Information Revolution: The Explosion of Features

The next major advance in visual instructions occurred during World War II, when the military used pictorial language to train soldiers. The Walt Disney Company, for instance, adapted its cartooning skills to create training documentation and films such as a movie employing Mickey Mouse to explain how to use a Browning .50 water-cooled machine gun. The defense industry in general also played a role in augmenting and applying visual instructional language during this period.

The visual techniques developed during World War II were quickly taken up for use in instructions for consumer products, although the necessity for these brilliant tricks has only become pressing recently. While many new items were introduced at the beginning of the 20th century, technology remained relatively simple: not many functions and a button for each one. Even during the Fifties and Sixties, irons, radios, ovens, cars and all the other blessings of modern society were still rather easy to use. Instructions in train stations, post offices, and airports were

This Room is Equipped With

Edison Electric Light

Do not attempt to light with match. Simply press switch on wall by the door.

The use of Electricity for lighting is in no way harmful to health, nor does it affect the soundness of sleep.

This sign was hung in New York City hotel rooms in the late 19th century.

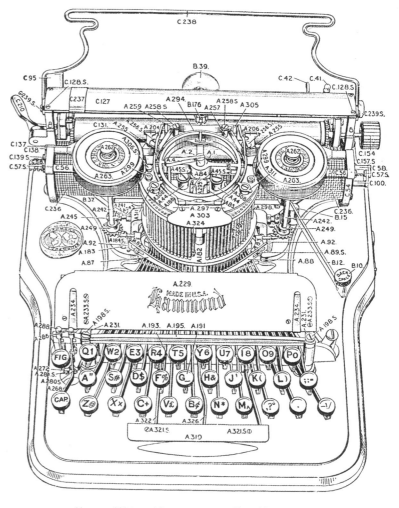

Hammond Universal Square typewriter. United States, 1913.

limited in number and mainly textual in form. The language of visual instructions was still in its infancy.

It was not until the end of the Sixties and the beginning of the Seventies that "featurism" began invading our products and that visual instructions began invading our lives in manuals and on packaging and products, through pre-set buttons, tone settings and safety lids. Since the early Seventies this featurism, in addition to the growth of the do-it-yourself industry, mail-order shopping and mass consumption of technological consumer products, has led to the more universal application of the language of visual instructions.

The real feast of the features only dates to the Eighties, when the microprocessor was introduced into our TVs, VCRs, microwave ovens, car stereos and cameras. Now the telephone is no longer just a telephone but a home switchboard with an answering machine and a fax with a 200-number memory, e-mail, repeat and follow-me functions. It's only since the Eighties that we've had to learn to program our own VCRs, pocket calculators, radios, microwave ovens, computers and electronic organizers. During the Eighties, ergonomists lost the battle with

This illustration from a Pfaff shoemaker's sewing machine manual (Germany, c. 1895) shows an arrow used to indicate the direction in which the wheel should be turned.

Dutch version of a Singer sewing machine manual. United States, 1920.

The illustrations along the top of the next four pages represent Philips radios produced between 1927 and 1997, with one example from every ten years. At first, the number of buttons increases; then it decreases, as each button came to be used for several functions. The complexity of use was not reduced, only hidden.

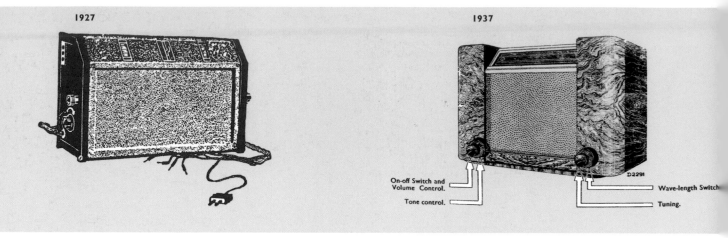

1927

1937

On-off Switch and Volume Control.

Tone control.

Wave-length Switch

Tuning.

D2291

From 1880 to 1963 Anker Steinbau produced expensive toy construction sets made of real stone. The blocks could be used to build model castles and bourgeois homes. After World War II, under the Communist regime, Anker Steinbau continued to make building blocks, but now they could be used to create factories and other buildings for the people. This illustration is taken from a pre-war manual.

The Art of Instructional Design

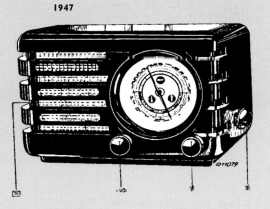

1947

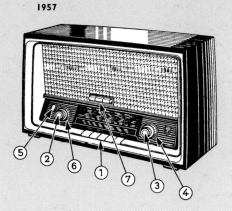

1957

electronics engineers, computer scientists and programmers to make these products intuitive to use. And we've passed the point of no return.

Engineers versus Designers

Ergonomists and user interface designers may try hard to make products instantly comprehensible to us, but they can't keep up with the electronics engineers and software programmers who develop more powerful chips and new features. As a result, the next generation of car stereos will have twice as many functions and will be twice as difficult to use. Aside from a few improvements added by the ergonomists and designers. And there's no reason to suppose that the situation will be getting any better. Let's look at an example to see how this affects those of us trying to use these ever more complicated products.

Until recently, programming a VCR meant punching buttons and numbers for channel, start and end time yourself. Then it became possible to use a bar code reader. To program, you slide the reader over codes printed in the manual for "Begin," "Hour," "Minutes," "Channel" and "End." This is only half a solution, of course. Now it's possible to print a small barcode next to every individual program. With such a guide, all you have to do is slide the reader over the code for a specific program. End of problem, right? Wrong. In the meantime, the number of channels has grown so enormously that it will soon be impossible to include them all in a single printed listings guide. And we will need new

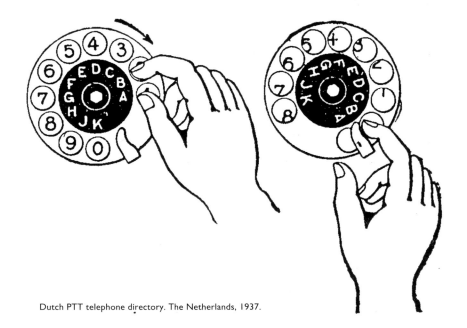

Dutch PTT telephone directory. The Netherlands, 1937.

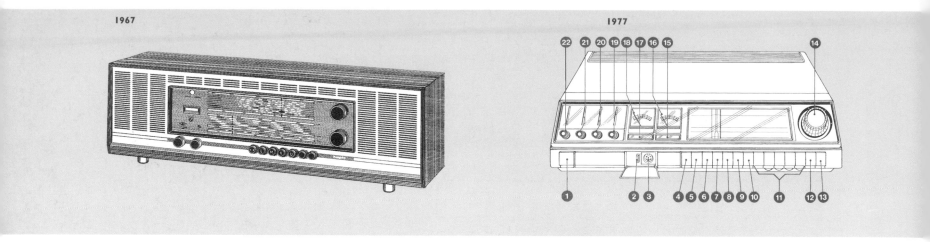

1967

1977

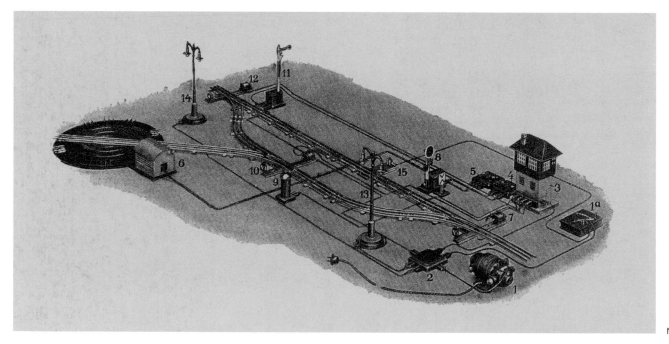

Märklin toy railway yard. Germany, c. 1930.

technologies to tell us what's on TV and how to program the VCR. Schedules are already provided on the Internet and programming can be done instantly. But this means even more options! Now you can type in the name of your favorite actress and immediately see when she'll be on and program your VCR accordingly. Technically possible today, generally available in a few years time. But how to explain this database query via the Internet to Joe Schmoe? That would take quite a few pages in a manual. Old problem solved, new problem created.

Märklin model Mercedes. Germany, 1936.

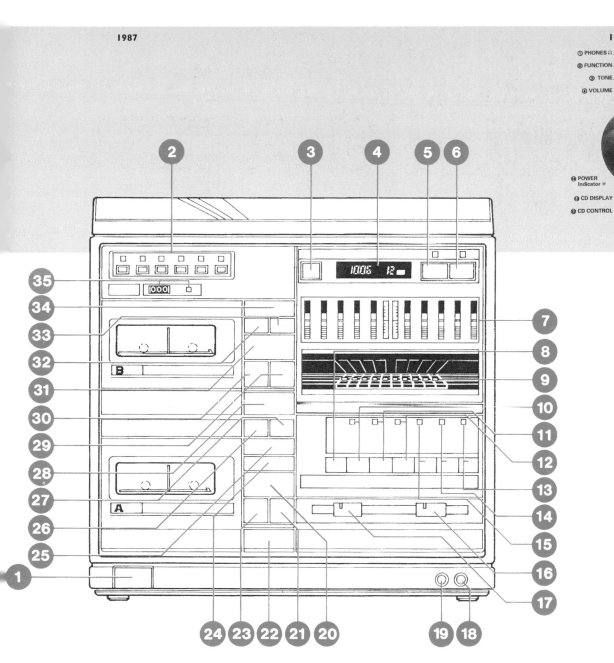

① PHONES ∩
② FUNCTION
③ TONE
④ VOLUME
⑤ CD DOOR
⑥ OPEN-CLOSE
⑩ FM-STEREO Indicator ✳
⑦ BAND
⑧ TUNING
⑨ TUNING DIAL SCALE
⑪ CASSETTE DECK
⑫ MIC
⑬ POWER Indicator ✳
⑭ CD DISPLAY
⑮ CD CONTROL
⑯ CASSETTE Compartment

Many other examples could be used to demonstrate that this process has only just begun. In four technological generations (four years' time), the manual for one Canon laser printer grew from about 48 pages to several books, in total some 750 pages. Moreover, new products come to market with dozens or even hundreds of features right from the start. The Canon Wordtank, an electronic dictionary that includes a datebook among its many features, comes with a 220-page manual (per language). This calculator-sized gadget is nearly dwarfed by its instruction booklet.

AEG washing machine manual.
West Germany, 1950s.

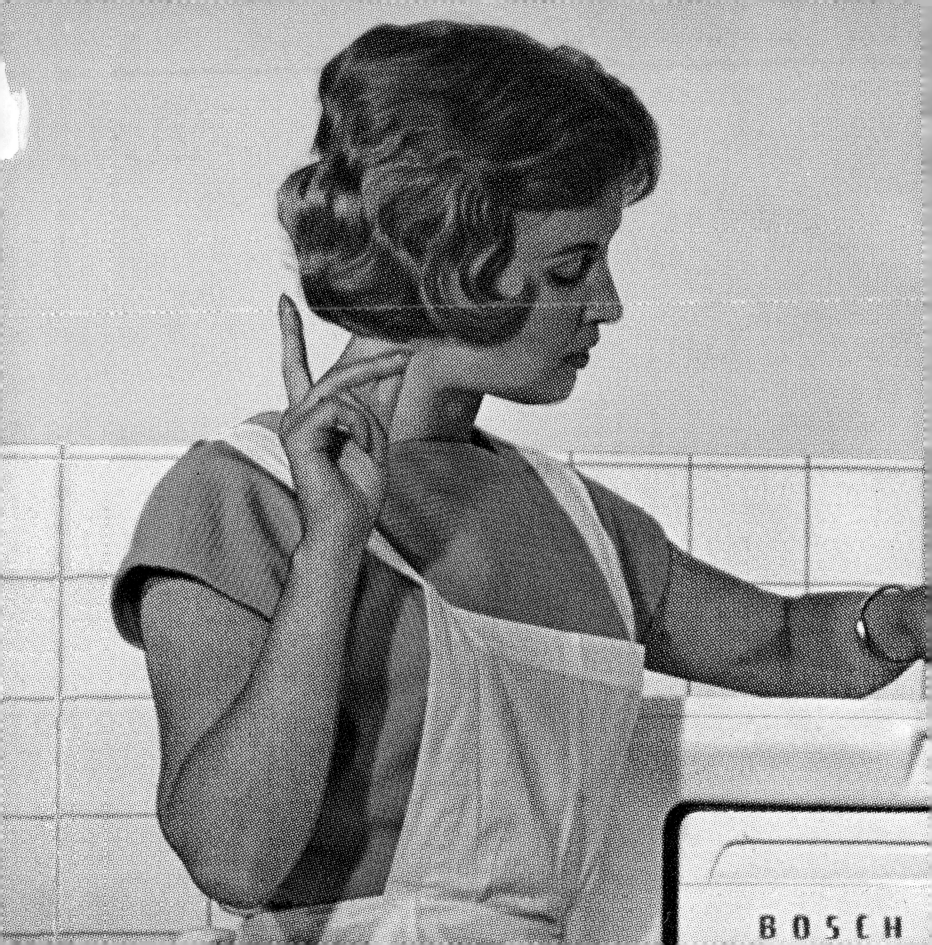

Feature Masochism

"So why don't we get machines with fewer features?" some people ask. Because they don't sell. In the early Nineties, Philips introduced Easy Line, a series of radios, TVs and VCRs with just the basic functions and no fantasy features to impress the neighbors. It goes without saying that these products were simpler to use and somewhat cheaper to buy. Well, it was still the early Nineties when Philips stopped producing Easy Line products—because nobody bought them. Don't think this was an exception. Ever heard of LetterPerfect? It's a word processor based on the famous WordPerfect program. When WordPerfect 5.0 was introduced, many people complained that it was loaded with functions they didn't need, that made the program difficult to use. So the company produced LetterPerfect 1.0, a word processor that could perform all the basic functions. It was also much easier to learn and much cheaper. But LetterPerfect never got further than version 1.0. Nobody bought it, either. A third example: Microsoft Write for the Macintosh. Same story: this stripped-down version died an early death, while its more complicated big brother, Word, amassed more and more features—and became the new world market leader. It seems masochistic, but we want those damn features, at any price.

Bosch refrigerator manual, 1960s. See page 51 for a selection from its text.

Volkswagen Bus manual.

West Germany, 1960

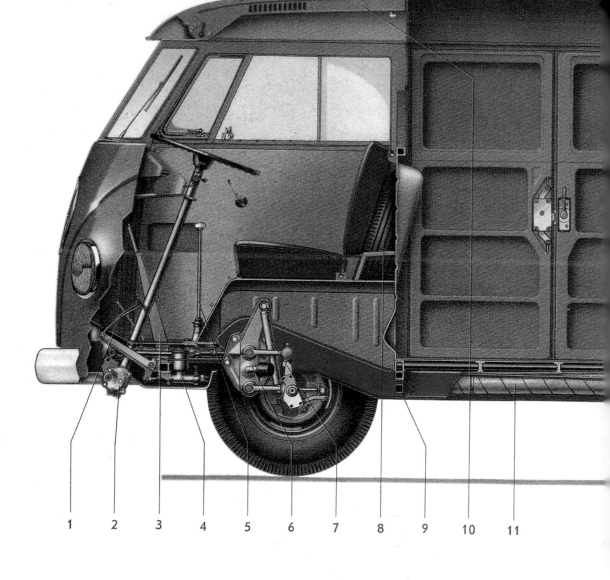

The illustrations along the bottom of the next four pages have been taken from Mercedes-Benz manuals between 1905 and 1991. The earliest images are technical drawings made by technical illustrators. Later, graphic artists created illustrations specifically for end users: the car buyers. Notice the evolution from overview illustrations with details indicated, to a focus on specific parts including instructional elements such as arrows.

1   2   3   4   5   6   7   8   9   10   11

1905

1927

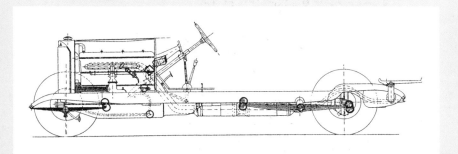

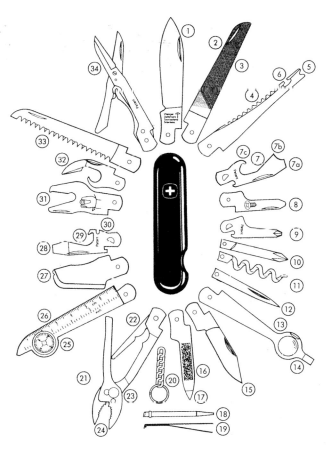

In 1886 the Swiss Army decided to equip every soldier with a regulation single-blade folding knife. In 1889 a new rifle was introduced. To disassemble this rifle a screwdriver was needed, so a multipurpose tool incorporating a knife, screwdriver, reamer and can opener was created: the Swiss Army knife. It wasn't until after World War II—when American G.I.s discovered this tool—that the Swiss Army knife entered into worldwide distribution. As with electronic products, the number of features has grown rapidly in recent years. This illustration is taken from a recent manual for a Wenger knife.

Instructional Hieroglyphs

How do we cope with these complicated products and features? We learn to read instructional hieroglyphs. Mass production, mass consumption, high-tech electronics, the do-it-yourself industry, the development of packaging (for almost everything) and the growth of international transportation and trade have urged the development of a universally comprehensible language: visual instructions, the imperative pictorial Esperanto of our time. This language has flourished in a century of visual entertainment, of movies, television, illustrated magazines, cartoons. Visual communication in general has triumphed over text. Pope Gregory the Great would at last be satisfied.

Since World War II, visual instructions have been developed for almost every type of product, from toys to computers, traffic signs to chopsticks. Such instructions have evolved from mainly text to mainly pictures, and

1934   1952

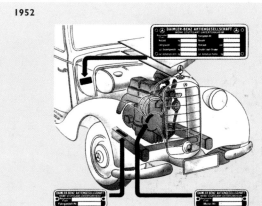

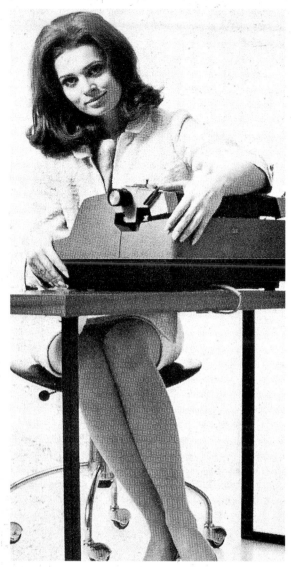

IBM 72 typewriter manual. United States.

these images have gone from being just pleasing or informative to functional. They have also progressed from general overview pictures to detailed instructional series. Furthermore, realistic drawings and photographs have been replaced by abstract symbols, icons and pictograms—the hieroglyphs of our time.

Many of the techniques applied in visual instructions have been adapted from cartoons: anthropomorphic drawings (sweating televisions, coughing printers), lines that indicate movement, depictions of sounds ("Click!"), balloons and inserts containing texts. Visual instructions have also developed a cartoon-like vocabulary of their own: metaphoric drawings, symbols and icons such as the wastebasket that means "delete file"—all in the effort to keep up with the newest technologies created by over-inventive engineers.

We don't need to fire our engineers, but we should put them under restraint. Technology is far too complicated to leave to them. Meanwhile, we have to go on opening bottles, jelly jars, milk cartons and cheese packets. We have to insert coins, ink cartridges and bank cards. We have to set up do-it-yourself furniture and assemble plastic toys for our kids. And with the introduction of even more futuristic features, explained only in the language of pictures, we'll need advanced technical training to get through daily life. Twenty-first century technofrenzy.

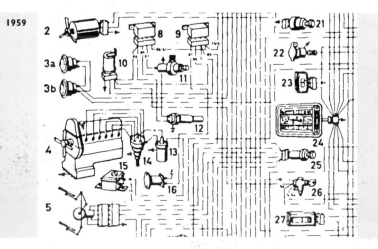

1959

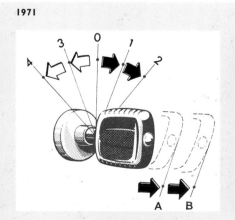

1971

1984

The Art of Instructional Design

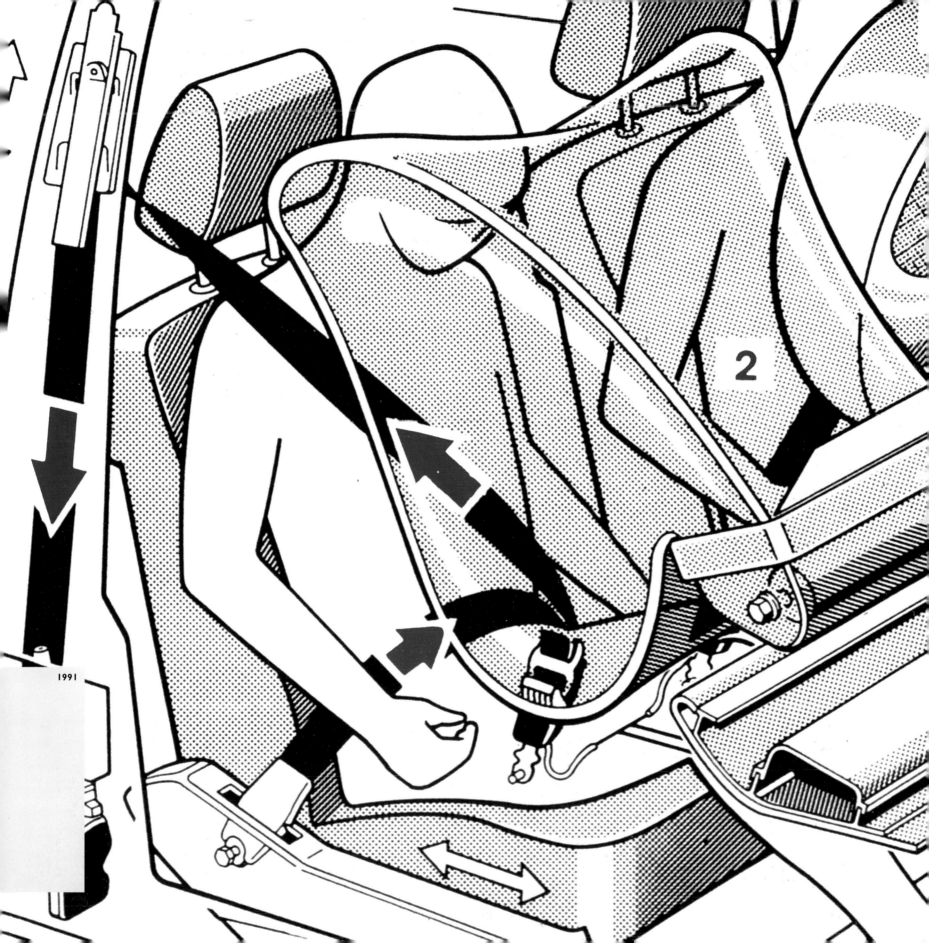

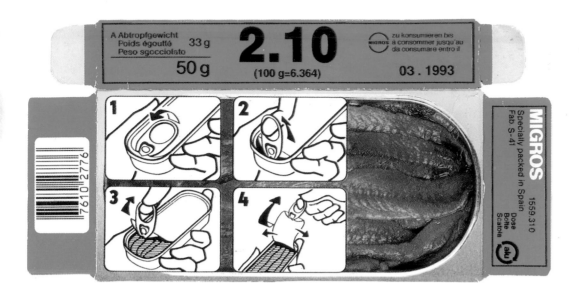

The Medium is the Message

Since the introduction of microprocessors, electronic consumer products have become black boxes.

Form Follows Function or Form Frustrates Function?
"Form follows function," the sculptor Horatio Greenough advised around 1850, and architect Louis Sullivan made the phrase famous in 1896. Content first; design comes later. But architects, engineers and designers rarely follow this rule. "Technology first! Design first! Expression first! Creativity first! Me first!"—that's how they think. As a result, we still have to ask ourselves: Where is the entrance? How do I open this door? What's the function of this button? Why is that little red light on? How do I put this call on hold?

Products are often advertised as being self-explanatory, but they're not. The medium is not the message. Not if the product itself is supposed to be the medium. That's why we need extra instructions. To open doors and packages. To unpack a microwave oven. And to program it. To transfer a telephone call and to insert a table in a text document. In fact, to do anything at all in high-tech life we have to follow instructions, because the medium is not the message.

The Semantics of the Button
On rare occasions the medium is almost the message. The ink cartridge of the HP DeskJet printer, for instance, incorporates is own message. The top of the cartridge takes the shape of an arrow that indicates in which direction to push it exactly into the correct position in the printer. Great design of an instruction. The medium here is the message—but it still comes with an instruction sheet needed to complete the installation of the cartridge.

Of course there are many other examples of products made in such a way that their design—the way they look or feel— informs you how to use them. A century ago, light switches had buttons that indicated whether the current could go through or not. If it could, the switch was in line with the electricity cable, which gave the visual impression that the current was flowing directly through the switch. If the current was blocked, the switch looked like a cross-bar. Buttons for cruise control are another example: the tactile impression provides information. In some cars, if you push a concave button, you go faster, and if you push a convex button, you reduce speed. (Or is it the other way around?)

Many other tricks have been employed to make products more user-friendly. Buttons for the operation of related functions are

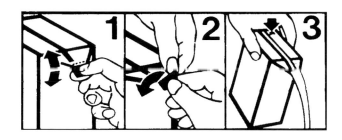

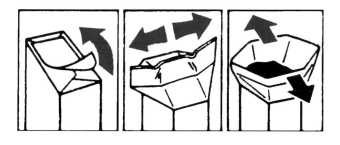

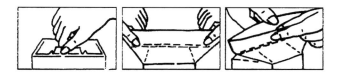

grouped together or have the same color. The graphical user interface (the computer screen that looks like a picture instead of merely lines of type) has been developed for some products. Such a user interface may seem more visual than verbal, but it's not. Most of the instructions are given in words rather than pictures, grouped together in menus, sub-menus, windows and fill-in forms. Only the icons for direct manipulation, such as the scissors that mean "cut text" in a word processing program, are truly graphical. And these icons in particular require explanation through a context-sensitive on-line help system.

Help All Over

So we get extra information to operate our products. Sometimes instructions for use are already included in advertisements. Most often instructions are printed right on the packaging. You can also find instructional graphics on buttons, near displays and on screens. Then there are labels on wires and stickers on covers. And then there is the manual ("Read this manual carefully before operation"), sometimes two or three. Plus a list with errata. Wall-charts accompany copiers, medicines come with package inserts for patient and doctor. Software may be included, or a videotape with instructions, demonstrations and an interactive tutorial. We also get wizards, personal assistants and guides to help us set up a PowerPoint presentation. In fact, we're completely bogged down with visual instructions.

Auditory, Tactile and Visual Messages

We perceive instructions with three of our senses: hearing, touch and sight. As far as we know, no one has developed instructions using smell and taste ("If you smell roses, press the red button. . . ."), but someone's undoubtedly working on it. Until then, instructions will be communicated through auditory, tactile and visual means only.

"For travel information, press 1. . . ."

Auditory commands are probably the most frequently used type of instruction. The shouting sergeant ("March! Dismiss!"), the whistling cop, the warning mother ("Be careful!"), the instructing teacher ("Write this down"). Auditory instructions can be spoken directly, amplified with a megaphone or loudspeaker, or recorded on tape or CD-ROM. Many products make sounds that instruct. The camera beeps: "Flash is ready."

However, most high-tech sound feedback means trouble: "Battery running low" or "Printer out of paper."

The Little Push in the Back

Tactile instructions can also be communicated using different mediums. A little push in the back ("Go that way") is a tactile instruction. So are bullets, grenades and atomic bombs: "Go away" they mean, or "Submit." The design of many products and tools incorporates tactile information, but this is not always instructional information. For example, the handlebar in some World War II planes for "Wheels in or out" looked and felt like a row of small wheels, whereas the handlebar for "Flaps up or down" looked and felt like small flaps, because young and

Not Just Humans

Some animals instruct each other visually. Bees, for instance, indicate the direction and distance to flowers by movements on their honeycombs. The speed of their turns indicates distance, while the direction of their movements communicates the path to the flowers.

inexperienced pilots too often pulled up the flaps instead of the wheels and vice versa. The shapes didn't tell them when or in which direction to push or pull, but helped them to distinguish between the controls for two types of equipment.

"Read this before. . . ."

Visual instructions include facial expressions, arm movements, waving flags, traffic lights, pictograms and icons and complete animated movies. The size of a visual instruction may vary from a tiny, simple arrow on a telephone card to the airplane hangar's worth of documentation needed to build a Boeing 747.

Man is the Medium

Many present-day printed visual instructions are based on bodily gestures and movements. We use our bodies to communicate a wide variety of visual instructions. The winking of an eye or the lifting of an eyebrow may signify "Keep quiet!" The movement of an arm or finger may command "Move" or "Come over here!" Many of the instructional gestures used by policemen, arbiters, umpires, conductors, choirmasters, ship captains and airplane marshalers have been standardized. The policeman shows the palm of his hand to say "Stop!" The conductor lifts his baton to instruct "Fortissimo." Some instructional gestures have become icons. The police officer's "Stop" lives on as the computer icon that signifies "Stop! If you continue, you may lose data!" Other gestures have been mechanized, such as the signaling arms used by railroads.

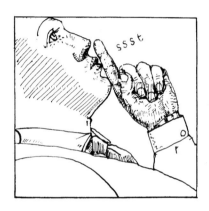

A White Wall is a Fool's Paper

The product may not always be the medium of the message, but the message can be transmitted using almost any medium. Smoke, flags, and light can all be used to instruct visually. But most visual instructions are printed—on anything that will accept printing. Arrows, lines, words and other visual

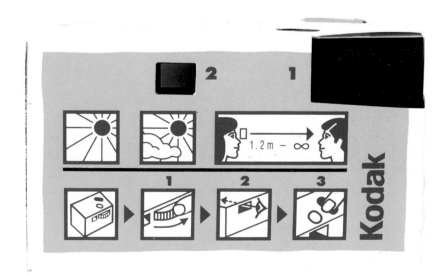

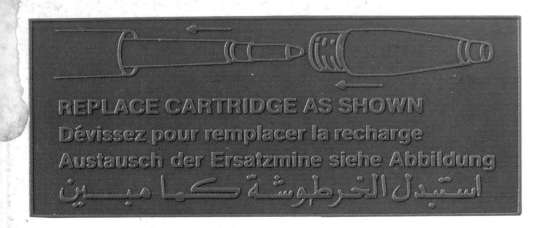

REPLACE CARTRIDGE AS SHOWN
Dévissez pour remplacer la recharge
Austausch der Ersatzmine siehe Abbildung
استبدل الخرطوشة كما مبين

Solid Litter Only | Cans

Cups | Leave Bag On Your Seat

No Cigarettes | No Drinks

Как свернуть вашу сигарету "DRUM"

1 Возьмите из пачки необходимое количество табака "DRUM" (около 1 гр)

2 Распределите табак равномерно по длине сигаретной бумаги "Mascotte" Проклеенная полоска

3 Начинайте осторожно двумя руками скручивать сигарету для достижения характерной формы

116.691 H

DRUM® Excellent CLASS A TOBACCO

DRUM-Excellent is een bijzondere melange, samengesteld uit 17 geselecteerde tabakssoorten, van goudgeel tot diep rijpbruin.

Deze soorten met hun eigen karakter en geur, geven aan DRUM-Excellent half zware shag, zo'n aparte, pittige en toch zachte smaak!

Vervaardigd en gegarandeerd door

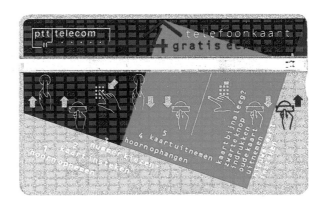

The Art of Instructional Design

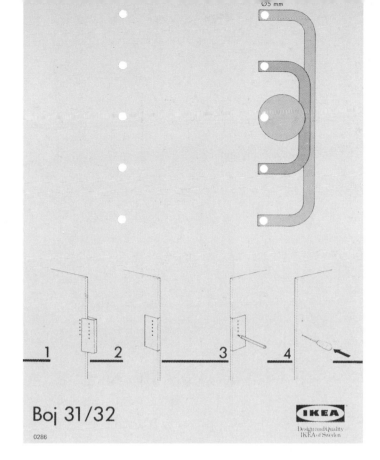

Boj 31/32

IKEA
Design and Quality
IKEA of Sweden

0286

instructions are printed on glass jars, iron lids, tins, plastic sheets, metal covers, felt-tip pens.

Some instructions—usually those for do-it-yourself products —include a template. You fold out the manual or sheet, place it over a piece of wood, draw lines, and drill holes or saw along the markings indicated. In this case, the medium is indeed the message.

Occasionally, the medium is deliberately *not* the message. Some cigarette lighters in the U.S. are now designed so that smokers must light up using two hands, just like you need two hands to use a chainsaw. And of course, smokers now need visual instructions to show them how to do this. To prevent accidents, lawyers and juries have decided.

OPENEN   SLUITEN

The Medium is the Message

N. 79 SEDANI — 10/12 min.

I. 86 FUSILLI — 7/9 min.

N. 88 RUOTE — 8/10 min.

N. 90 CRESTE DI GALLO — 8/10 min.

N. 85 PIPE RIGATE — 8/10 min.

N. 84R PIPETTE — 8/10 min.

N. 83 ABISSINI — 8/10 min.

How Should I Tell You?

"When I use a word," Humpty Dumpty said in rather a scornful tone, "it means just what I choose it to mean—neither more nor less."
"The question is," said Alice, "whether you can make words mean different things."
"The question is," said Humpty Dumpty, "which is to be master—that's all."

Lewis Carroll, *Through the Looking Glass* (1871)

Variation in Concepts

Many roads lead to Rome. Designers of visual instructions must not only choose from a variety of mediums, but also from a wide range of concepts, illustration types and details. The right concept is especially important if instructions are to be comprehensible. What will be explained in text, and what in pictures? How will pictures and text be combined? What kind of pictures will be used? An exploded-view summary drawing with reference numbers in picture and text? Or a series of detailed instructional pictures along with captions in balloons? A series of overview drawings with numbered instructional texts around them? Or simply a column of text on the left and a column of pictures on the right? While the possibilities are infinite, the right choice must be made, for the concept behind a visual instruction can strongly influence the user, just as in advertising.

For products of the same type, instructional messages are usually more or less alike. These products have similar functions and fairly standardized user interfaces. If you know one VCR, you know them all. That's why the people selling basic electronic equipment know so well how it works, not because they're so much smarter than us. Yet, although the content of instructional messages for such products is quite similar, that content is usually presented in very different ways. That's why you have to learn how a manual is designed before you can start using the information in it. For simple instructions, such as "How to insert a telephone card," inconsistency in instructional design leads to simple problems—a few

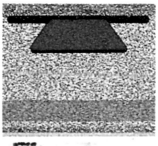

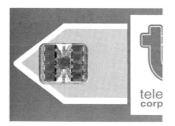

How to insert a telephone card. Three ways to do it wrong. Dozens of ways for designers to tell us how to do it correctly.

seconds wasted. But try a more complex task, such as buying a subway ticket from a vending machine for the first time—you cannot immediately start pressing buttons. You must find out what you have to find out and how to find that out, Or try to figure out what to do in case of water landing by reading the safety card of a foreign airline. And then there's the infamous VCR. You probably will want to program it at some point, and you won't find it easy to do. Let's have a closer look at these three levels of complexity in instructions for use.

How to Insert a Telephone Card

For a telephone card, the message is not much more than "Insert this way." If you can't find the instruction or fail to get it right the first time, simply insert the card the other way around—or upside down or sideways. One arrow is enough. But each telephone card has a different method for informing us to "Insert this way." Various types of arrows, or a pointing finger, or a hand, or a text with an arrow or an arrow with a text, or a text in an arrow. Moreover, you never know where the instruction will be printed: left, right, middle, top, bottom. Any place seems right—for the designer.

What to Do in an Emergency in a Plane

For passengers, all commercial airplanes are more or less the same from an instructional point of view. Whether it's a Boeing or an Airbus, the same situations may occur and the same questions have to be answered. How to sit during an emergency landing. How to open the emergency exits. How to use seat belt, oxygen mask and flotation device. Plus some information about stowing luggage, smoking bans, etc. Now look at the safety cards: in our collection of more than 100 "Do not remove from aircraft" cards, no two are alike.*

*Exception? The cards from Aeroflot, the airline of the former Soviet Union, all seemed exactly equal. Maybe some are more equal than others, but we didn't find any differences.

First of all, the messages are not the same. Airlines apparently disagree about what exactly they need to tell us. Some show how to inflate boats, others don't. Some tell us to crawl across the floor in case of fire, others don't mention fires at all. Next, even when the cards present similar information, the method of presenting it differs completely, though all the cards are almost totally pictorial. The order in which things are explained varies from one card to another, too. How to sit during an emergency may be first, or where to leave the plane or what not to do. The care of children may take up the most space, or the emergency exits or how to reach them. You might expect the cards to present instructional steps in the order that you should perform them, but this is certainly not the rule. And then there's the visual concept, the way in which pictures (and text, when used) are combined. There may be a central overview illustration, or a series of small instructional pictures, or a combination. The pictures may be presented from left to right or in columns. The only thing airplane instruction cards really make clear is that the number of ways to express something in visual language may be even greater than in written language.

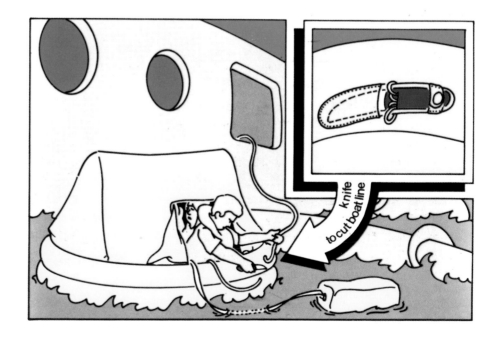

knife to cut boat line

The Art of Instructional Design

In case of water landing, women should care for babies and men should go hunting for food.

According to this image, at least.

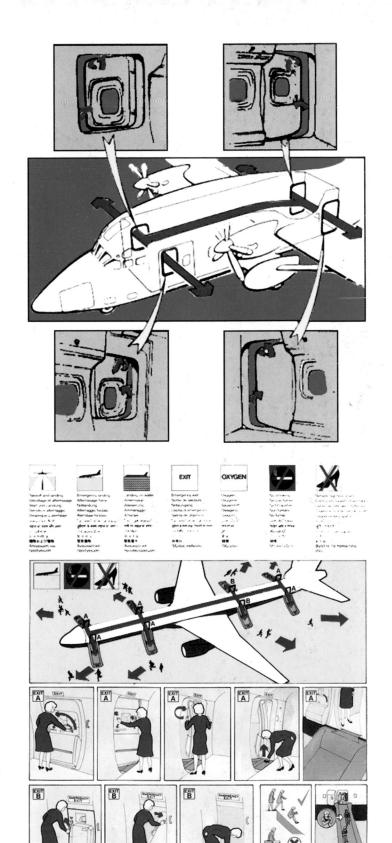

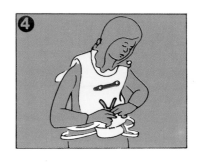

The Art of Instructional Design

How to Program the VCR

But who cares about airline safety cards? Nobody reads them anyway. It's different if you want to program your VCR. Now you will need the manual. A few users will start by reading it, but most will go straight to the machine, picking up the instructions only when they encounter a problem. The manual has to be user-friendly to both types of people, not an easy task. And manufacturers and designers appear to have many different ideas about how to instruct us— or it seems they have no idea at all, but simply want to express themselves visually. So we get a totally new manual for each new VCR—sometimes totally different manuals for VCRs from the same company at the same moment of production. Consider just these three, each presented through a completely distinct concept:

- A wire-bound manual with pages that can be folded out and placed on top of the VCR for easy reference. Each page features an actual-size photograph of the display, with red lines connecting details to short instructional texts.
- A fold-out cover filled with small pictures and a separate book with text referring to the pictures.
- An overview drawing with numbers plus instructional texts combined with detailed instructional pictures.

This variation in concepts indicates how wide-ranging our ideas about good visual communication can be, and not only for designers. Ask 100 regular people what a camera manual should look like and you'll get at least 100 totally different answers. There is only one thing they'll all agree on: the manual they got with their camera is totally wrong.

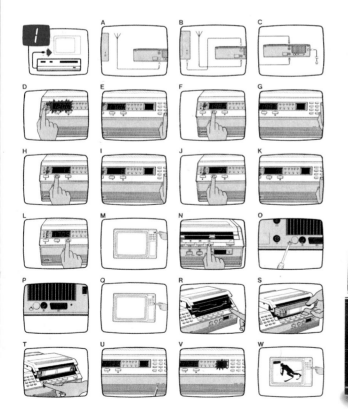

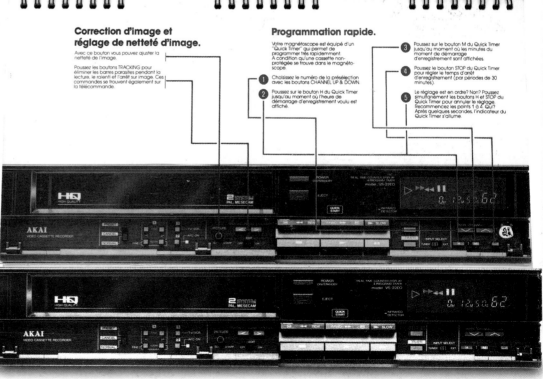

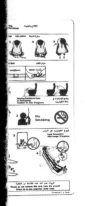

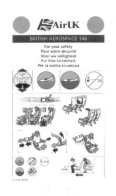

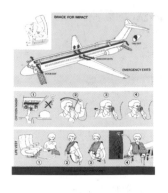

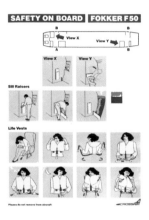

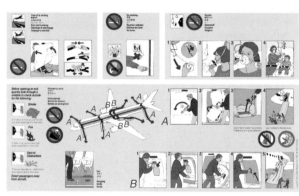

Standardization

Still, there is hope. Slowly, very slowly, we'll get used to a Standardized Generalized Symbol Language (SGSL). We already recognize the visual Esperanto for *paper jam* ●⚙️↯, *print darker* °▪°°♔°, *cut* --✂--, *copy & paste* 🖼️ 🖼️, for *turn the page* ✍️ , *loudness up* +◁ and *loudness down* −◁ and similar expressions. This simple visual language originates from technical professions such as mechanical engineering, electronics, and sea and air traffic control, which have a high degree of standardization in visual communication. In the aircraft industry, for instance, the International Air Traffic Authority has compiled big books to standardize the order in which instructions should be presented, what technical illustrations and drawings should look like, how language should be used, which icons, pictograms and symbols are appropriate, and so forth. Surgeons have another standardized dialect. So have electronic engineers.

Some of these conventions are spreading to the rest of us via electronic and mechanical consumer products and international travel. Committees have standardized thousands of symbols, with more and more of them destined for the instruction of non-specialist users. But how many of these symbols do we recognize? Traffic signs have been officially standardized for some time, so they're understood worldwide. But that's about it, even though there are many more symbols to learn. For example, the icons for on/off buttons have been standardized. That is to say, there is a standard icon for *on* | and one for *off* ○; one for *stand-by* ⏻, one for *start* ◇ and one for *stop* ▽; one for *emergency stop* ⬤ and one for *fast stop* ▼; one for *on/off with one button* ①, one for *on/off with a push-button* ⊕, one for *on* ⊙ and one for *off* Ȯ for a specific part of the equipment. . . .

The consequences of all this? Out of the thousands and thousands of icons and symbols we are confronted with, we only recognize 300 or so. Take a look around your house, office, public transportation system. Can you describe or draw the symbol for

grilling of flat food items ▢ on your microwave oven? The symbols for *contrast* ◕ or *brightness* ☼ on your remote control? The symbol for *balance* ◺◿ on your car stereo? The *mute* ⍾ icon on your telephone? The *Entrance* ⇥ and *Exit* 🏃 icons in subway or train stations?

To make the problem more complicated, in most cases no one is required to use the official standard even if there is one, and graphic designers can be very clever at inventing new dialects. The official sign for "No smoking" depicts a cigarette in a barred circle. But every airport, every hospital, every cinema, every town hall and every post office in the world seems to display its own visual instruction for "No smoking."

The meaning of some icons may be easy to guess, but the majority are not, at least not to most of us. Designers struggle between iconicity and ideogrammaticality. Should an icon look like the object it stands for or should it be a more symbolic representation? The traffic sign for railway crossing portrays a steam locomotive 🚂, which is rarely seen anymore, so the symbol does not really look like the train as we now know it and has a low degree of iconicity. But the steam locomotive has become the archetype of the train, so it is highly ideogrammatic and therefore understandable to most people.

Internationalization and technological developments may eventually force us to learn this *extra* language, and it's not an easy one. We have already learned verbal language, which is very powerful, especially due to its grammar, the rules that allow us to infinitely vary a limited set of words. Comparatively speaking, visual language has very little grammar. New

USE OF LIFE RAFT

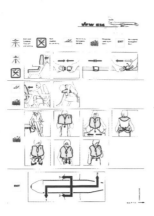

VFW 614

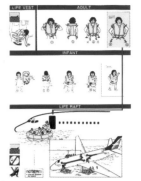

LIFE VEST · ADULT · INFANT · LIFE RAFT

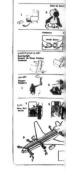

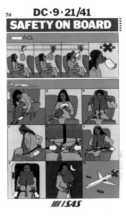

Do not inflate inside the aircraft

DC·9·21/41
SAFETY ON BOARD

747 LR SUD · SAFETY INSTRUCTIONS

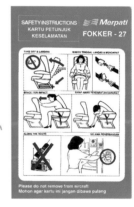

SAFETY INSTRUCTIONS
KARTU PETUNJUK
KESELAMATAN
Merpati
FOKKER - 27

Please do not remove from aircraft
Mohon agar kartu ini jangan dibawa pulang

LIFE JACKETS

AIR FRANCE
BOEING 737
Consignes de sécurité / Safety instructions
Sicherheitsvorschriften
Consignas de seguridad / 安全のしおり

AIR FRANCE
BOEING 737
Consignes de sécurité / Safety instructions
Sicherheitsvorschriften
Consignas de seguridad / 安全のしおり

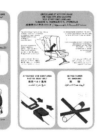

YOUR EMERGENCY EXITS · UW NOODUITGANGEN · IHRE NOTAUSGÄNGEN

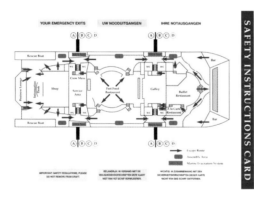

SAFETY INSTRUCTIONS CARD

Evacuation on water

Adult life-jacket

Adult life-jacket for child

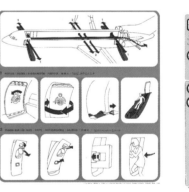

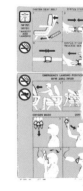

British Caledonian   BAC 1-11 (500)

Safety instructions
Consignes de sécurité
Instrucciones de seguridad Instruções de segurança

BAC 1-11 Do not remove from aircraft
Ne pas enlever de l'avion Favor no retirar del avion
(500)   Nao retirar do avião

EXIT · SORTIE

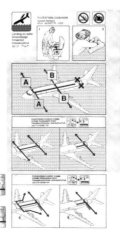

FLOTATION CUSHION

SIKKERHED OM BORD
MAERSK AIR · FOKKER 50

SAFETY   DC-9
ON BOARD   FINNAIR
TURVAOHJEET · SÄKERHETSINSTRUKTIONER

FASTEN SEAT BELT · NO SMOKING

AT TAKE-OFF AND LANDING · OXYGEN

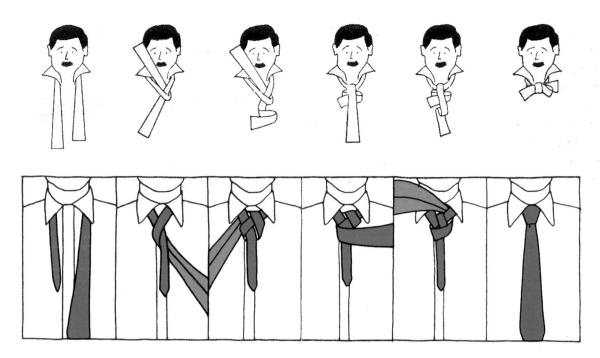

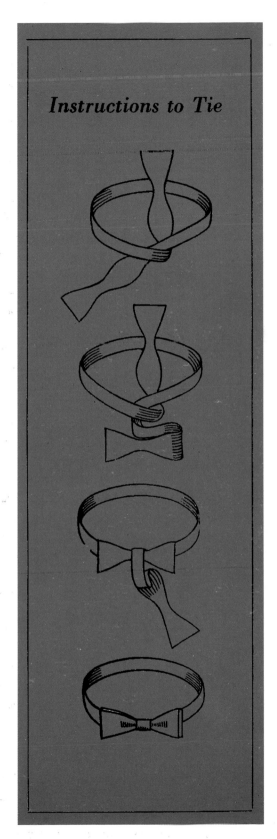

grammar may be evolving, however. For example, > means *forward* and >> means *fast forward*. Could this be the beginning of a grammatical rule for visual language: doubling an icon always means more, faster, bigger? Maybe some day. But until more rules are established, designers will tell us the same things in visually different ways. And we'll have to reconstruct what they mean, over and over again. Until standardization is more universal, only technicians among themselves will speak a common visual language. Engineerology.

Who would want standardized directions for tie-tying anyway? But presenting the steps as they'd be seen in the mirror is a clever idea.

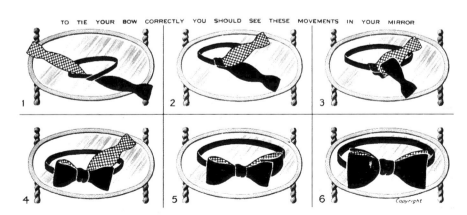

TO TIE YOUR BOW CORRECTLY YOU SHOULD SEE THESE MOVEMENTS IN YOUR MIRROR

A Manual for Each of Us?

He thinks that the customs of his tribe are the laws of nature.
George Bernard Shaw, *Caesar and Cleopatra* (1898)

For Christian, Jew, Muslim and Hindu; For Chinese, American, Aboriginal and European

Visual instructions must be universally comprehensible. For the upstate New Yorker and the rural Chinese visiting New York City for the first time. For the 16-year-old German whiz-kid and the 78-year-old French technophobe. For Christian, Jew, Hindu and Muslim. Everybody must be able to understand the instructions and nobody should be offended. So you find a lot of cultural differences in instructions for use, right?

Wrong. In a collection of more than 30,000 manuals we could find no clear cultural differences. Even instructions for the use of condoms are the same for Hindus, Muslims, Christians and heathens. Europe's largest producer of condoms, which sells its products worldwide, does not produce different instructions for different cultures. If the instructions are universally identical for such a delicate matter as condoms, why would they be different for more ordinary products?

The English and Swedish versions of the Volvo 480 manual include a drawing which shows how a pregnant woman should put on her seat belt. Yet the Arabic version omits this drawing. A cultural difference? Maybe so. But why then is the drawing also missing from the French version of the

 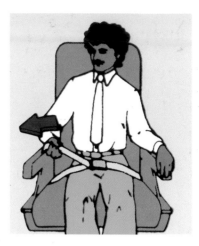

manual? Do we see here a distinction between Anglo-Swedish and Franco-Arabic cultures? A bit unlikely. In fact, the image was left out of the Arabic and French versions of the manual because these languages require more page space than English and Swedish.

Of course, there are differences. In the safety card of Air Maroc, for example, we see drawings of people who look Moroccan using the emergency doors. In the Kenya Air safety card we see black people and in Swissair safety cards we see white people. In the Japanese manual for an Olympus camera we see a Japanese woman showing how to hold the product and in German instructions for leaving a building in case of fire, we see a white man. But these can hardly be called *cultural* differences. Therefore, we can conclude only one thing: there are no different visual instructions for different cultures.

Hao Yu Usim Kloset

Pidgin English instructions from New Guinea depicting "How You Use [Water] Closet."

Dear Madam!

You are now the happy owner of a Bosch refrigerator. We are pleased that you have chosen our product.

We know that the traditional "user manuals" are not liked very much, especially not by the ladies. Be honest. Did you read the manual of your sewing machine or your washing machine? No? Well, there you have it!

Perhaps you are now thinking: what does a man know about domestic affairs? But the man who is writing these lines really does know something about it. Of course he owns a Bosch refrigerator himself and now he wants to tell you about all the experiences that he and especially his wife have had with this refrigerator. These experiences are for the benefit of all women who own a Bosch refrigerator. In this sense we request you to consider the following explanations and now, for your benefit, listen to what the expert has to say.

Cleaning of the Bosch refrigerator

"Now that really is the limit," my wife said, when she read this heading. "You of all people want to teach us, housewives, what cleanliness is."

"Please, let's not get personal, darling," I replied, "I am not trying to tell housewives they should clean their refrigerators (because that is obvious), but to tell them what the best way is to do so."

"Now you tell me," my wife said, "How come these rubber linings are greasy?"

"Because you touch them with your own hands, dearie. I have noticed you have the habit of opening the door just slightly using the handle and then opening it further by pushing against the linings. It would be a great help if you were to unlearn that habit."

(. . .)

"By the way," I continued, "I would like to advise you to arrange the chilling-space more efficiently."

"Excuse me, but this way is practical."

"It may be practical for you," I admitted, "but it is not efficient. Of course you can arrange your refrigerator any way you want, but do bear in mind that there are areas with different temperatures in the refrigerator and that you have provisions that need to be stored somewhat colder and provisions that need to be stored somewhat less cold."

From a Bosch refrigerator manual, 1960s. For an illustration from this manual, see pages 28-29.

Visual Instructions are a kind of a cartoon strip for adults. They often consist of a series of anthropomorphic drawings and product details. Bright colors indicate "look here," arrows tell us to "twist this way," jagged balloons surrounding "Click" let us know what sound we should hear when we "connect A to B," and splash-lines draw attention to what the finished product should look like. A childish language, that's for sure; too childish for children. Kids, of course, don't need it. They've grown up knowing how to work the VCR and the computer. They read cartoons that are genuinely funny, or at least intentionally funny. Children enjoy Disney cartoons; we struggle through our Roy Lichtenstein instructions. That's our pleasure—at best.

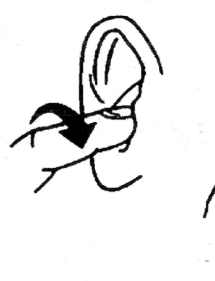

For Young and Old, Male and Female, Techies and the Rest of Us

On a sub-cultural level, though, we sometimes do see differences in visual instructions. Occasionally instructions for children are different from those for adults, instructions for men look different from those for women and instructions for techies may be different from those for the rest of us.

Young and Old

The "my first Sony" radio is clearly designed for children—although it appears quite popular with older people because it has such nice big buttons! Sony's world-band radio is clearly designed for adults, techies in this case. The design of buttons and icons often makes it evident for whom a product is meant. This is not only true for the product itself, but also for its instructions. The directions for LEGO construction blocks and for Technical LEGO are obviously different from those for a hearing aid. The LEGO instructions are full color, bright and bold, totally visual; no verbs. In contrast, the hearing aid manual is mainly textual—in a type size that's far too small for the elderly—and includes only a few, tiny illustrations. Of course, these differences in instructions may be due to the nature of the products. Perhaps LEGO construction blocks are better explained using pictures and hearing aids are better explained using verbs.

Instructions for use

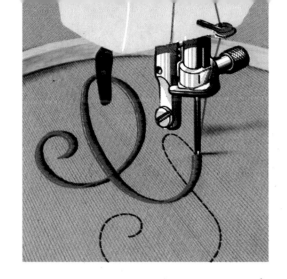

Male and Female

The Husqvarna chainsaw manual has a sturdy, masculine look with black and-white techno-style line drawings. The text commands, military-style: "Watch out!" "Hold this." "Press that." The manual for a Husqvarna sewing machine, however, looks like a women's magazine, with soft-focus photography, pastel colors and a pleasantly-toned text with many tips. The differences in presentation can only be explained by the different target groups for these products. Another example: the instruction manual for the Philips shaver varies visually from that of the Philips Cellesse cellulite massage system. Layout, pictures, phrasing and even typography are applied in a divergent manner.

It's bad enough that you have to remember to take a pill every single day, but do you have to be naked, too?

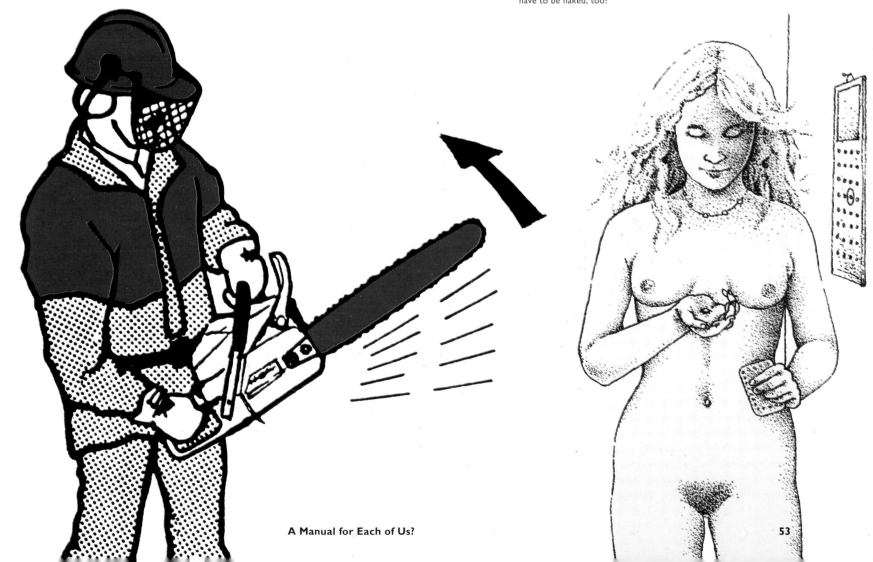

A Manual for Each of Us?

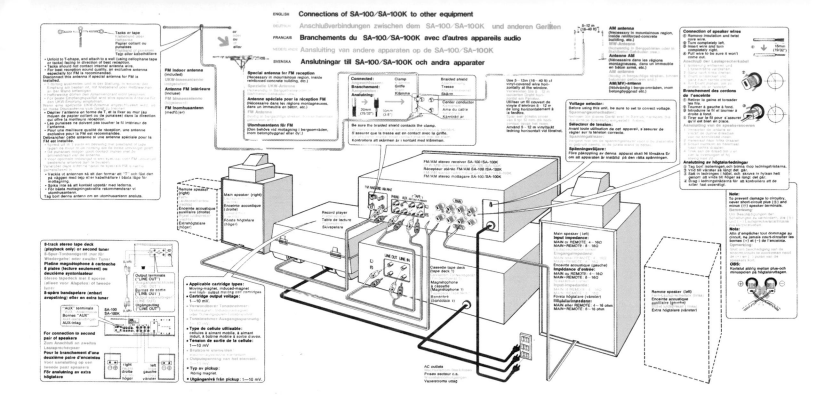

Techies and the Rest of Us

Engineers employ highly standardized methods of visualization. You have to be taught how to understand their drawings. Three-dimensional technical schemas are usually represented in axonometric drawings, with lines that do not converge into a vanishing point but stay parallel along one, two or three (isometric, dimetric or trimetric) axes. This type of imagery has advantages for specialists: it's easy to draw and makes it possible to calculate the real distance between two points by measuring the distance between them in the drawing and multiplying by a scaling factor. Professional technical illustrations also include many standardized symbols, especially in the electronics field.

The rest of us cannot easily understand this highly specialized visual language. But we often need instructions for highly technical products, so the specialists have created simplified variations. Instruction manuals for electronic products such as stereos include overview illustrations indicating how to connect speakers, aerial, amplifier, tuner, CD player and other components in realistic, not

To show illiterate miners in South Africa how to keep mine tracks free of rocks, this pictorial message was posted. But some miners read the instructions from right to left—and obligingly dumped their stones on the track. An urban legend?
From Henry Dreyfuss, *Symbol Sourcebook*, McGraw-Hill Book Company, 1972, p. 79.

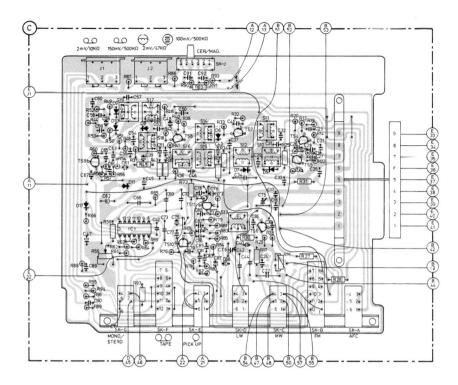

symbolic, semi-technical drawings. These are usually not axonometric but full perspective drawings, with lines converging into a vanishing point. This looks more like the actual wiring and cabling in front of us than the concepts that are illustrated in true technical drawings. In addition, symbols are not used, only icons (the more realistic little pictures), so that we can easily understand what is meant.

Usually though. . . .

Usually though, visual instructions are not designed for specific groups, not even when the products they explain clearly are. An electronic sketch pad, obviously intended for children (with bright colors and rounded corners), comes with a manual that looks like a notarial act, designed by an elderly engineer. This is the rule, not the exception.

If humans displayed some of the idiosyncrasies of current computer-user interfaces. . . . **Lon Barfield** compared our communication with a computer to the communication between a gentleman (us) and a butler (the computer). Hemelsworth is a rich, landed gentleman. Barker, his butler-to-be, is way ahead of his time. This passage concerns error messages.

Hemelsworth is waiting for his new butler to arrive. There is a knock at the door. Hemelsworth opens it.
Hemelsworth: *Hello.*
Barker: *Good morning. I'm your new butler, sir.*
Hemelsworth: *Oh, jolly good. I was rather wondering when you would arrive. What's your name?*
Barker: *Error code 3, sir.*
Hemelsworth: *That's an interesting name, "Error code 3," what shall I call you for short?*
Barker: *Error code 145, sir.*
Hemelsworth: *But that's longer than "Error code 3." Wait a minute, "Barker" was the name, wasn't it?*
Barker: *Yes sir.*
Hemelsworth: *Well come in Barker, come in. Would you like to join me for a cup of tea first?*
Barker: *Error code 19, sir.*
Hemelsworth: *What is all this "error code" business?*
Barker: *They're error messages sir. Here.* [He presents Hemelsworth with a book] *You look up the error message in here and it tells you what you did wrong.*
Hemelsworth: *I'm not doing anything wrong am I? And why do I have to look it up in a book? Why can't you just tell me straight out?*
Barker: *Error code 65, sir.*
Hemelsworth: *Error code 65? Alright let's give this a try then . . .* [He leafs through the book] *Hmmm . . . "Error code 65: too many questions at once." Yes, well. Come in, let's have a cup of tea. I think I'm going to need one.*

Lon Barfield, *The User Interface, Concepts & Design*, Addison-Wesley, 1993, pp. 26-27.

Eureka: Ingenious Elements

Ironically, the development of computers has led to a recovery and redeployment of this prehistoric form of communication. The use of icons is both ancient and current.

William Horton, *The Icon Book* (1990)

Creativity per Square Inch

No matter how good the concept, it's the creativity per square inch that makes the difference between great communication and unbelievable annoyance. So let's consider the details and phrasing of visual instructions. Where to look, what to do, when to act, what should happen, or what absolutely *not* to touch and when *not* to press . . . this is the stuff visual instructions are made of. The elements designed to express such messages are often ingenious inventions. These are the clever solutions of countless "Unfamous Artists."

From Over-emphatic Warnings to Dubious Results

In this section of the book, we will follow the order in which a designer should think and communicate with the user when creating visual instructions. *Warnings* come first—usually because the lawyers say so. Of course, it's useful to know what not to do from the beginning. How not to open the box and immediately damage your new radio. Then it's time to use your new vacuum cleaner, or construct that model

airplane. To do so, first we have to *identify* the pieces (and figure out what got left out of the box). We might need to know *measurements*, especially if we have to assemble something (is that the $3/4''$ or the $7/8''$ screw?). Next we have to look at the *composition*—to see what each piece is for and where it should go. We need to know the exact *location* and *orientation* of the parts to be assembled or connected. Let's hope the designer has found a clear way to tell us in what order we should do things, the *sequence* of our acts. If we're lucky, the illustrator skillfully used arrows, hands, dotted lines and other tricks to show us the *movements* and *connections* we have to make. Then, at last, it's time for *action!*

Of course, we'd like to know what the results of our actions should be, so we can compare what's supposed to happen with what we actually get: *cause and effect*. The manual says we should hear "Click!" but we actually hear "Crack!"—things like that. And finally, we hope to see a picture of *what it should look like*—but probably doesn't.

At times these categories overlap, and sometimes an element provides more than one type of information. A drawing identifying parts may also show how to connect them. A sequence of acts might end with an image of the desired result.

In the following chapters we present a number of creative, funny, ordinary and extraordinary examples for each of these details of visual instruction. Or simply the most beautiful ones—not necessarily the most effective.

> **Trials never end, of course. Unhappiness and misfortune are bound to occur as long as people live, but there is a feeling now, that was not here before, and is not just on the surface of things, but penetrates all the way through: We've won it. It's going to get better now. You can sort of tell these things.**
>
> Robert M. Pirsig, *Zen and the Art of Motorcycle Maintenance* (1974)

Warnings

If I do not know that I do not know, I think I know.
If I do not know that I know, I think I do not know.
R. D. Laing

No smoking. Do not touch the blades. Mind the roller, it's hot. Take care you do not. . . . *The imperatives.*

What do they look like in pictures? A drawing of the object with a red X crossing through it; a red bar; a prohibitory sign or a hand giving a stop sign. The exclamation mark to say: "Be careful," "Don't" or "Mind." There's not much more to express in warnings. Sometimes a combination of good and bad works better: "Do not do it this way" (canceled with an X) and "Do it this way" (with a check mark √).

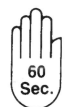

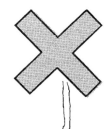

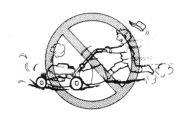

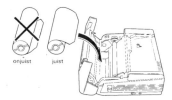

onjuist   juist

- Never
- Jamais
- Nie
- No
- 磁性面は
  さわらない。

 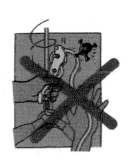

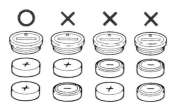

Destroy			
Tilintegør			
Förstöra			
Hävitä			
A Détruire			
Distruggere			
Vernichten			
Vernietigen			
Destruyase			

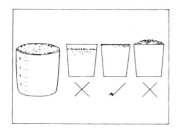

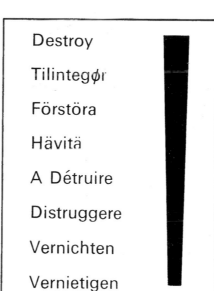

ATTENZIONE

This way   Not this way

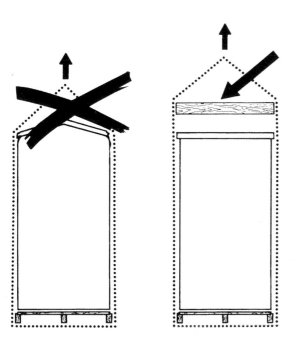

Plus de 40°C

Anthropomorphic drawings are used more and more to warn us about what not to do. Coughing copiers, frightened cameras, happy and sad batteries and stereos holding umbrellas. Sometimes in combination with X's, bars and check marks.

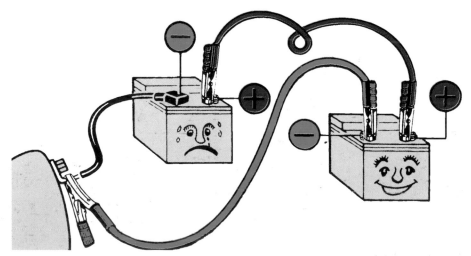

The Art of Instructional Design

Finally, never be afraid to try a feature!
You cannot hurt your VCR by trying any
of the features in this manual!
From a Magnavox VCR manual, which
nevertheless comes with a page-length insert
of cautions and warnings

Identification

That must be wonderful; I have no idea what it means.

Molière

Notches, connecting plates, DC OUT, the latch, the battery compartment, the EJECT button, the socket. *The nouns.*

Which batteries, which buttons, what to insert. . . . We have to identify and recognize the objects we'll be working with before we can work with them. "Which part goes here?" "What does the handle look like?"

Sometimes nothing more than a boring list of parts, sometimes a nice pictorial overview. Often nothing at all, leaving you to guess—"Would this be the 68-pins white scsi-cable plug?"

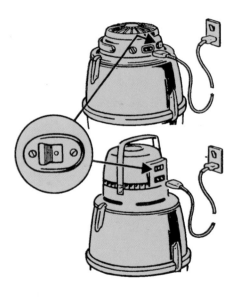

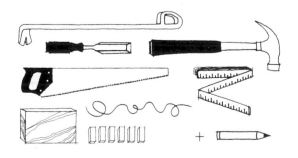

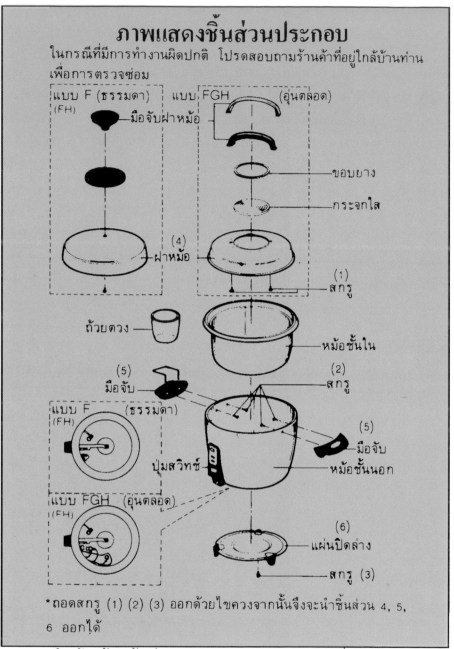

The Art of Instructional Design

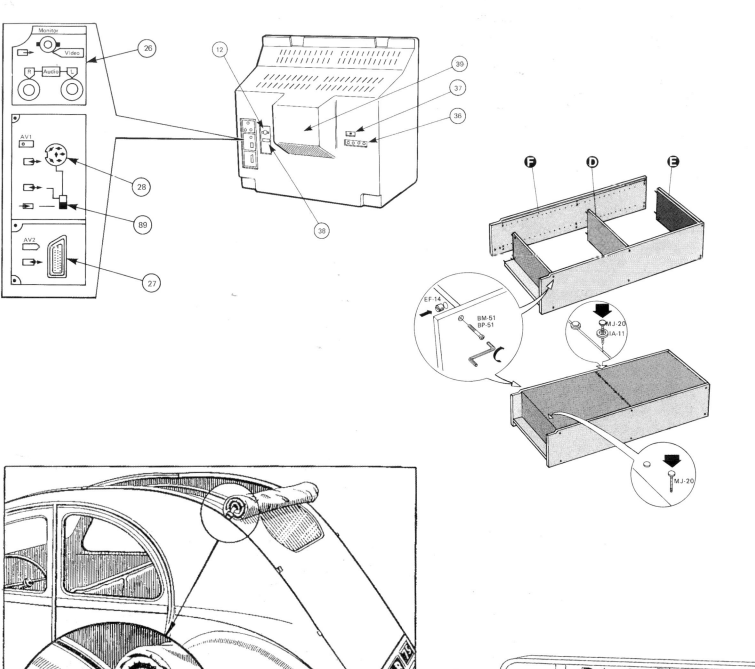

4

5

6

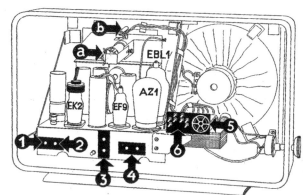

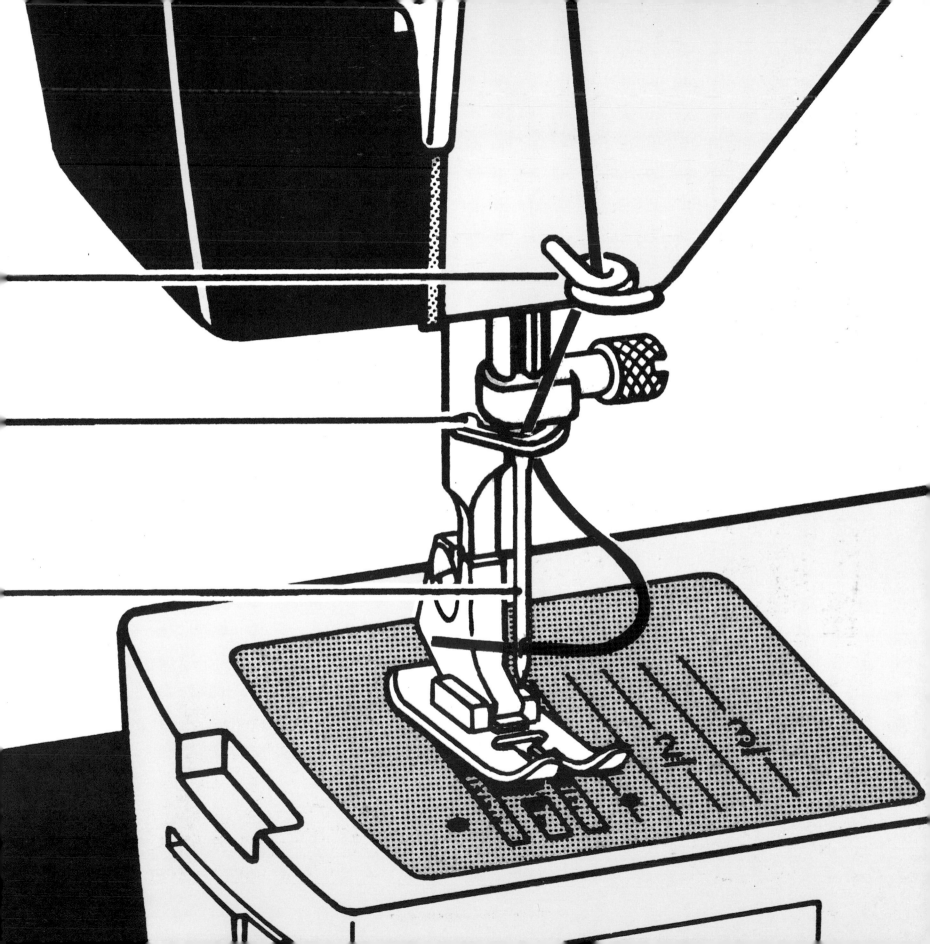

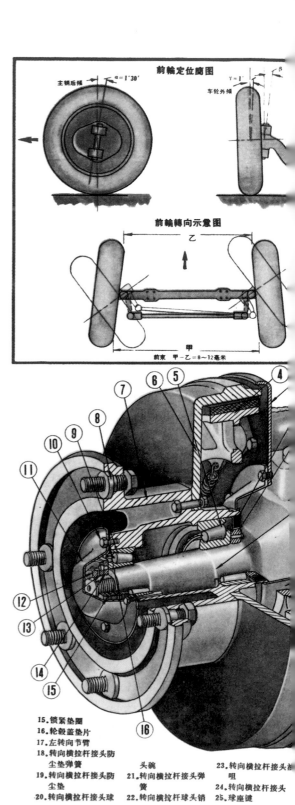

前輪定位圖

前輪轉向示意圖

前束 甲－乙＝8～12毫米

22

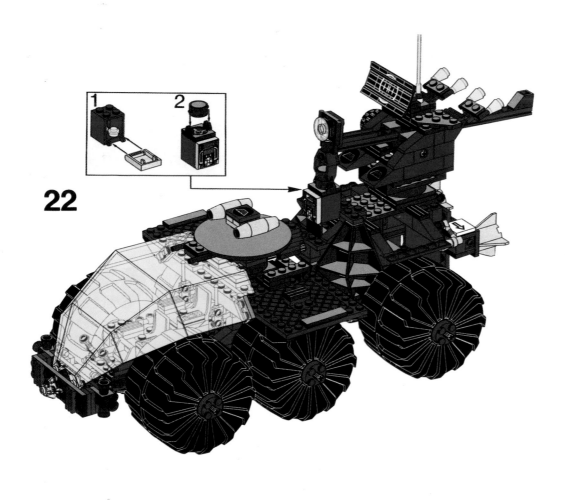

15.锁紧垫圈
16.轮毂盖垫片
17.左转向节臂
18.转向横拉杆接头防
  尘垫弹簧　　　　　头碗
19.转向横拉杆接头防　21.转向横拉杆接头弹
  尘垫　　　　　　　　簧
20.转向横拉杆接头球　22.转向横拉杆球头销

23.转向横拉杆接头油
  咀
24.转向横拉杆接头
25.球座键

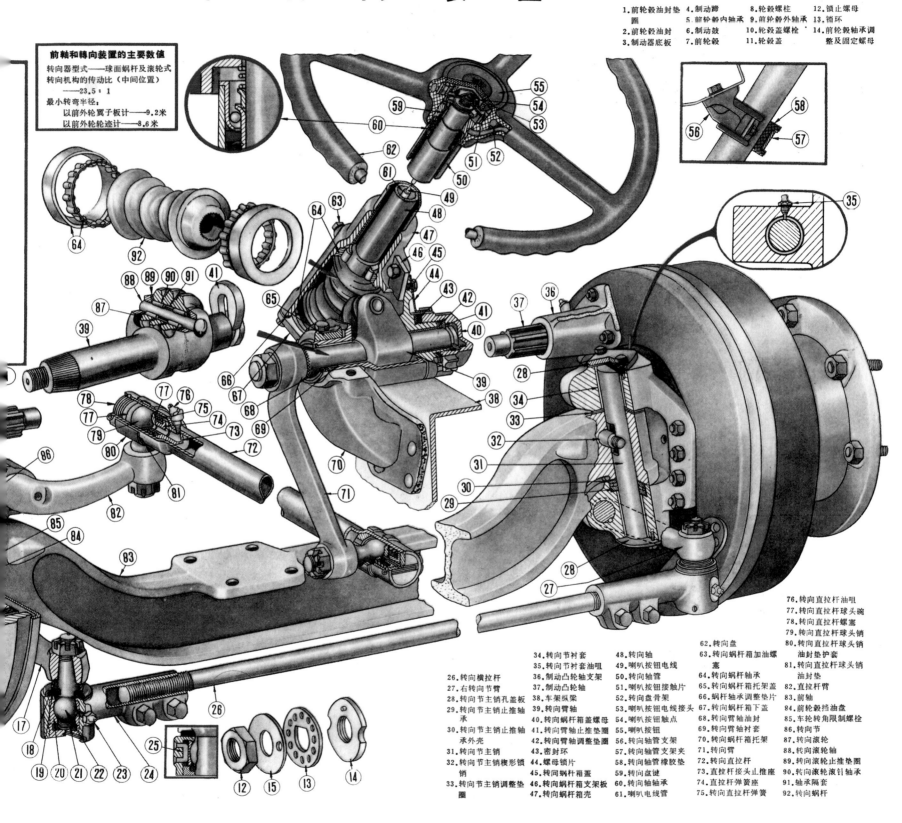

前 轴 和 转 向 装 置

前轴和转向装置的主要数值

转向器型式——球面蜗杆及滚轮式
转向机构的传动比（中间位置）
——23.5：1
最小转弯半径：
以前外轮翼子板计——9.2米
以前外轮轮迹计——8.6米

1.前轮毂油封垫圈　　4.制动蹄　　8.轮毂螺柱　　12.锁止螺母
2.前轮毂油封　　　　5.前轮毂内轴承　9.前轮毂外轴承　13.锁环
3.制动器底板　　　　6.制动鼓　　10.轮毂盖螺栓　14.前轮毂轴承调整及固定螺母
　　　　　　　　　　7.前轮毂　　11.轮毂盖

26.转向横拉杆
27.右转向节臂
28.转向节主销孔盖板
29.转向节主销止推轴承
30.转向节主销止推轴承外壳
31.转向节主销
32.转向节主销楔形锁销
33.转向节主销调整垫圈

34.转向节衬套
35.转向节衬套油咀
36.制动凸轮轴支架
37.制动凸轮轴
38.车架纵梁
39.转向臂轴
40.转向蜗杆箱盖螺母
41.转向臂轴止推垫圈
42.转向臂轴调整垫圈
43.密封环
44.螺母锁片
45.转向蜗杆箱盖
46.转向蜗杆箱支架板
47.转向蜗杆箱壳

48.转向轴
49.喇叭按钮电线
50.转向轴管
51.喇叭按钮接触片
52.转向盘骨架
53.喇叭按钮电线接头
54.喇叭按钮触点
55.喇叭按钮
56.转向轴管支架
57.转向轴管支架夹
58.转向轴管橡胶垫
59.转向盘键
60.转向轴管轴承
61.喇叭电线管

62.转向盘
63.转向蜗杆箱加油螺塞
64.转向蜗杆轴承
65.转向蜗杆箱托架架
66.蜗杆轴承调整垫片
67.转向蜗杆箱下盖
68.转向臂轴油封
69.转向臂轴衬套
70.转向蜗杆箱托架
71.转向臂
72.转向直拉杆
73.直拉杆接头止推座
74.直拉杆弹簧座
75.转向直拉杆弹簧

76.转向直拉杆油咀
77.转向直拉杆球头碗
78.转向直拉杆螺塞
79.转向直拉杆球头销
80.转向直拉杆球头销油封垫护套
81.转向直拉杆球头销油封垫
82.直拉杆臂
83.前轴
84.前轮毂挡油盘
85.车轮转角限制螺栓
86.转向节
87.转向滚轮
88.转向滚轮轴
89.转向滚轮止推垫圈
90.转向滚轮滚钉轴承
91.轴承隔套
92.转向蜗杆

Emergency Light Source | For Reading | Auto Repairing

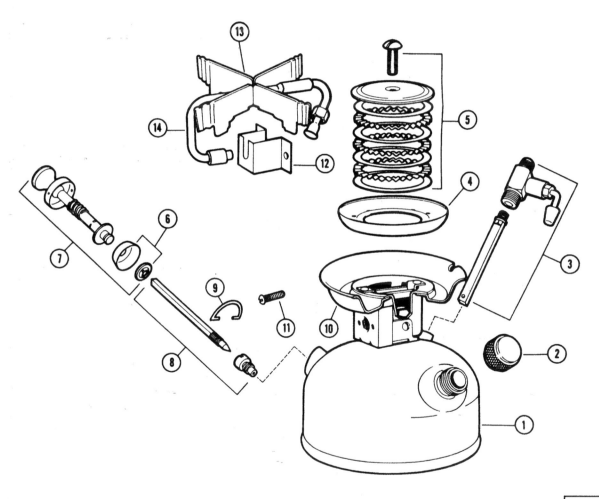

Identify where to look, what you should have found in the box and what else you need to make it work. You may find a reference system, with numbers explained in the text or in a separate list (usually textual but sometimes visual); zoom frames with details; hands, fingers or arrows indicating the elements in an exploded view or cutaway drawing; highlighted elements in photographs or drawings; and cutaways or ghost views to give you a general idea of the location of components.

DAIMON ALKALINE
DAIMON N1
DAIMON N2

■ = hervorragend geeignet

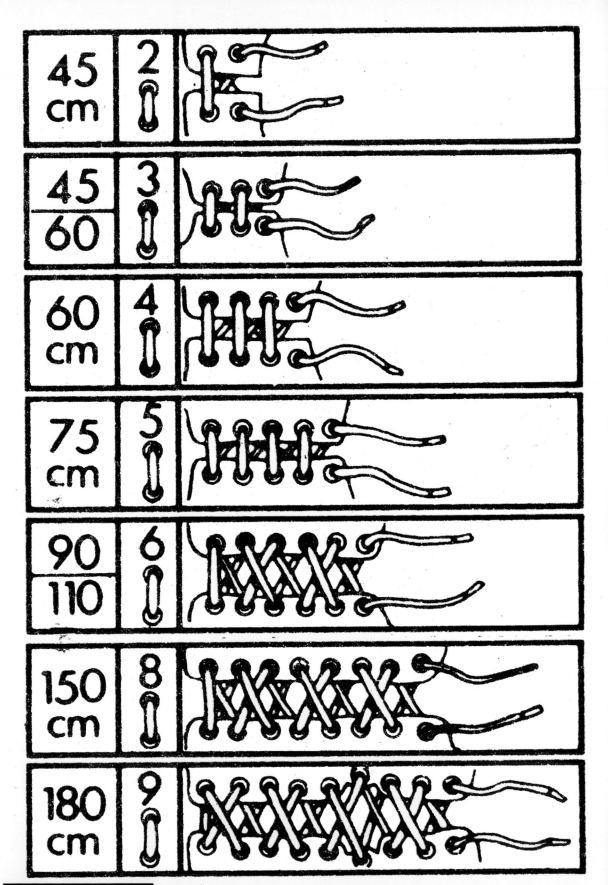

45 cm	2	
45 / 60	3	
60 cm	4	
75 cm	5	
90 / 110	6	
150 cm	8	
180 cm	9	

The Art of Instructional Design

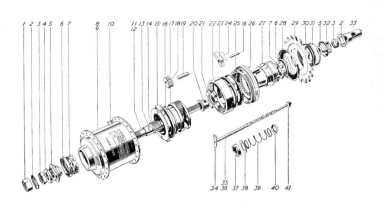

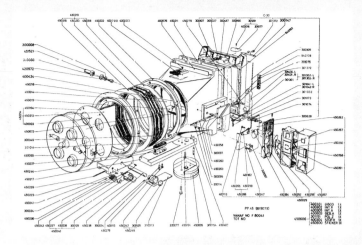

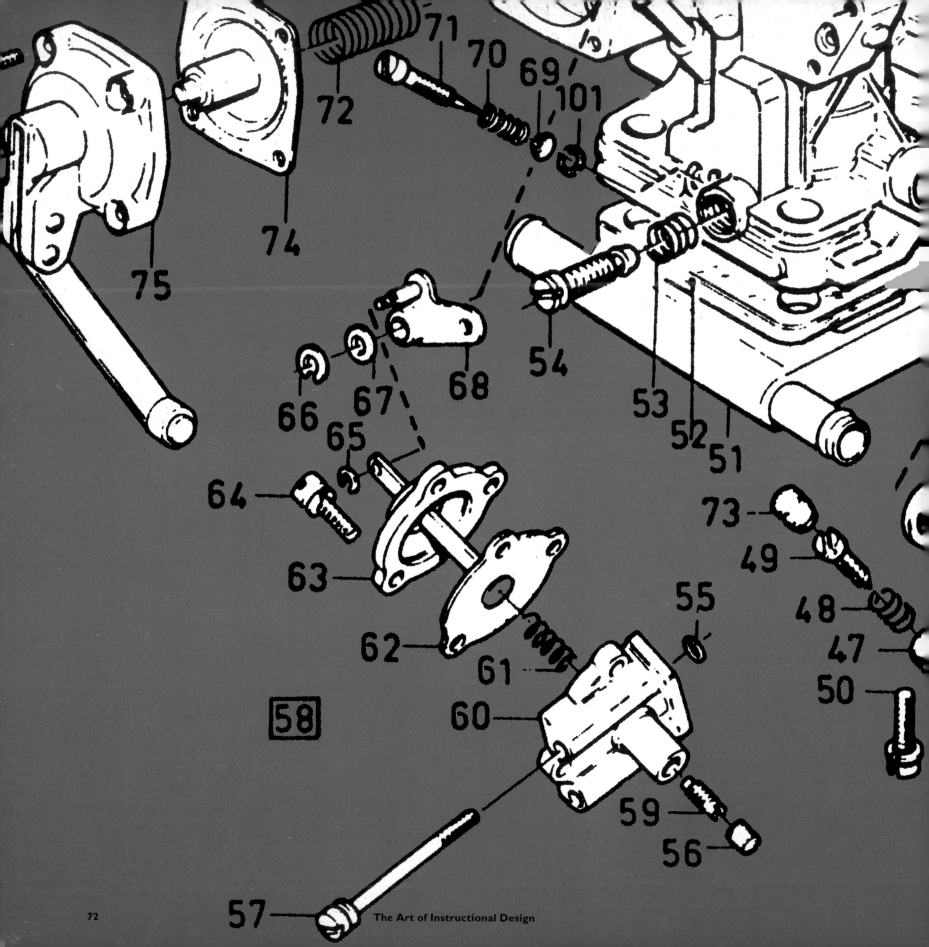

The Art of Instructional Design

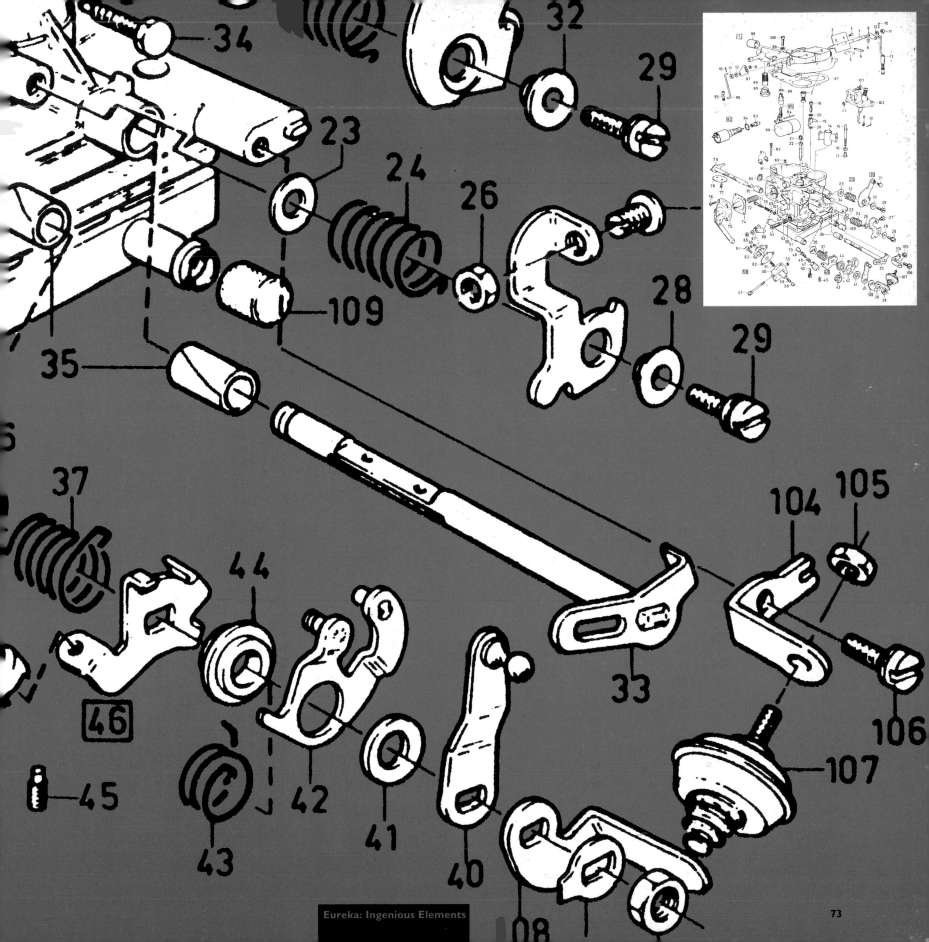

34

32

29

23

24

26

109

35

28

29

37

104

105

44

46

33

45

43

42

41

40

106

107

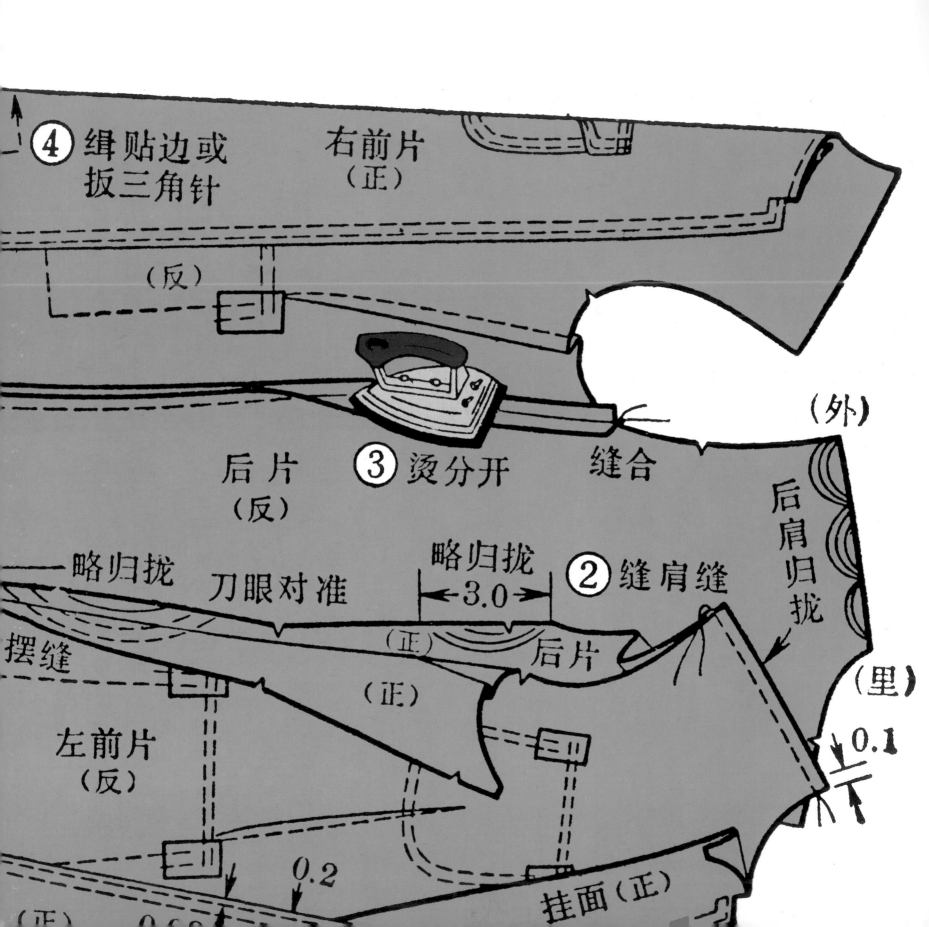

④ 缉贴边或
扳三角针

右前片
（正）

（反）

（外）

后片
（反）

③ 烫分开

缝合

后肩归拢

略归拢

略归拢
3.0

刀眼对准

② 缝肩缝

摆缝

（正）

（正）

后片

（里）

左前片
（反）

0.1

0.2

挂面（正）

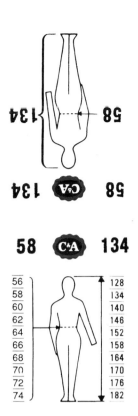

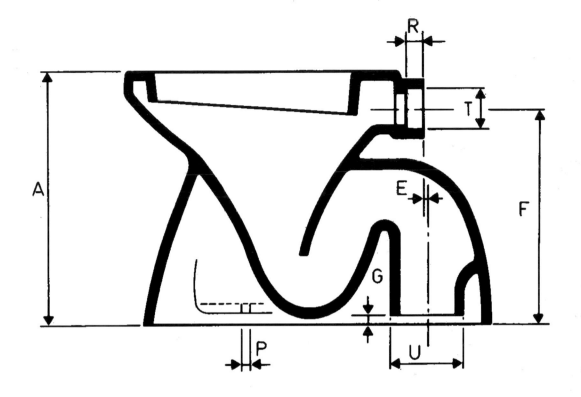

Measurements

Perfection is not reached when there is nothing to be added anymore, but when there is nothing to be left out anymore.

Antoine de Saint-Exupéry

"What should the distance from the corner be?" "At what length do I saw off the laths?" "How many ounces of water?" "Do not place more than 25 pounds. . . ." "The self-timer will release the shutter after 10 seconds." *The place and time adverbs.*

Millimeters or inches between arrows. Two clocks showing a difference of five minutes. Little weights with indications. Dimensions of time and place in simple visuals.

Sometimes a manual doubles as a template— then the manual really is the message.

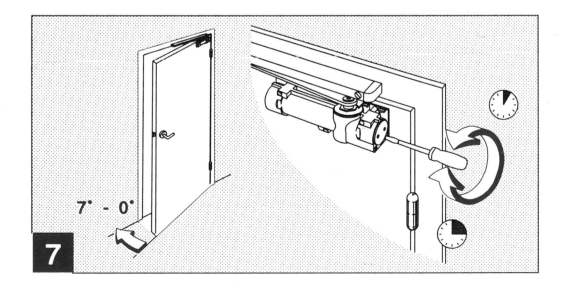

MAX

8

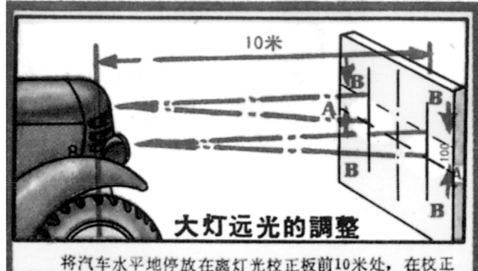

大灯远光的調整

　　将汽车水平地停放在离灯光校正板前10米处，在校正板上比大灯中心线低100毫米处划一水平线 A-A，根据大灯的中心划出中间的垂直线和两边的垂直线 B-B，打开远光，调整大灯的位置使灯光中心线与A-A、B-B线交点重合。

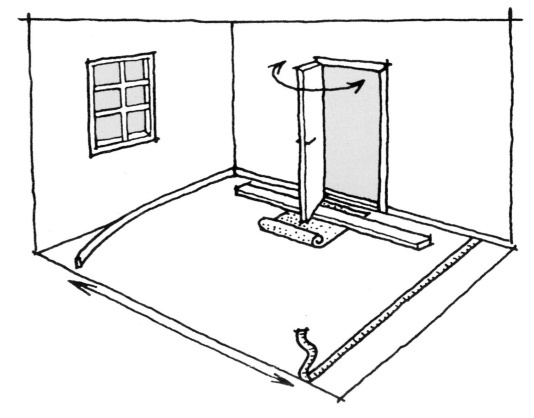

The Art of Instructional Design

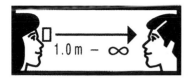

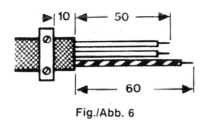

Fig./Abb. 6

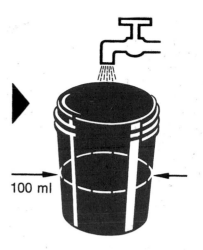

100 ml

Measurements most often inform us about sizes and distances, but sometimes also about quantities, weight and time. Size and distance are usually indicated using the standard two-headed arrow labeled with the distance between its ends.

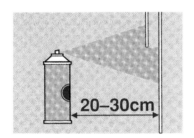

20–30cm

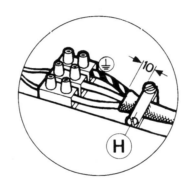

H

max 40 kg

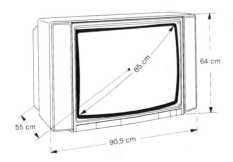

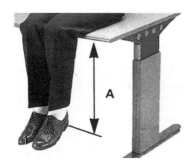

A

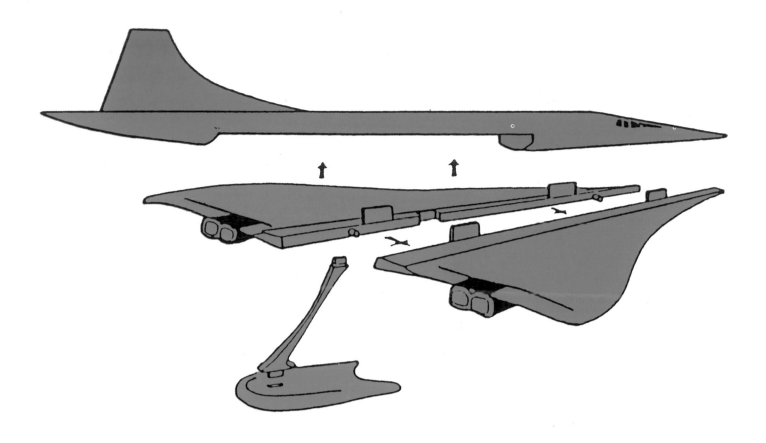

Composition

The world is complex, dynamic, multidimensional; the paper is static, flat. How are we to represent the rich visual world of experience and measurement on mere flatland?

Edward R. Tufte, *Envisioning Information* (1990)

"How does A relate to B?" "Where should that plug fit in?" "Where does this screw go?" "Plug it in backwards?" "The window should face toward. . . ." How will all the elements be combined? *More adverbs of place.*

This is the time for nice exploded views. For semi-professional, full-perspective illustrations. For the big picture.

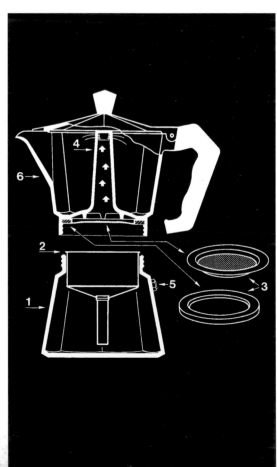

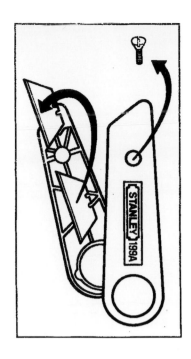

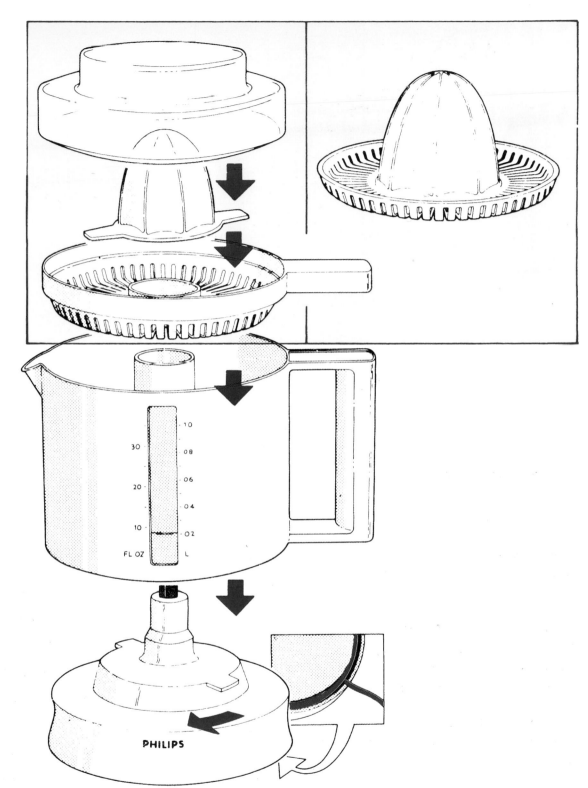

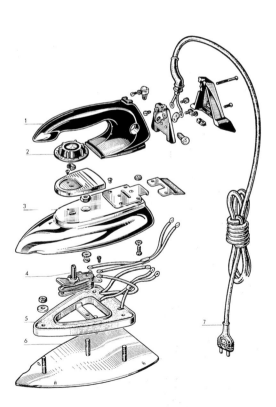

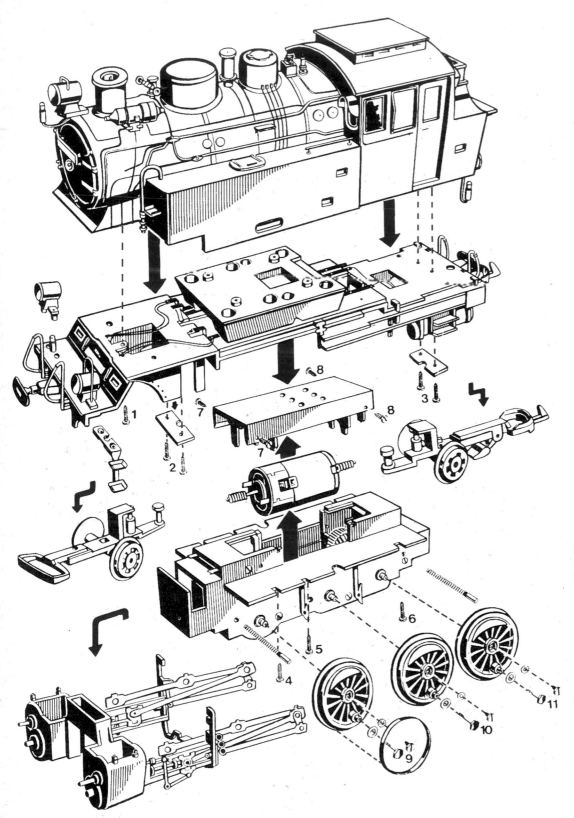

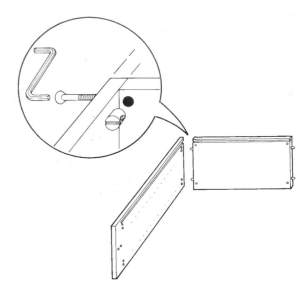

Nobody knows who invented the exploded view; maybe it was Leonardo da Vinci. Whoever did got it right—no one's improved upon it yet. There are two systems for exploded view drawings: isometric and full perspective. In isometric drawings, lines are parallel along one, two or three axes. In full perspective drawings, the lines seem to vanish at a point in the distance. This seems more realistic to readers without technical training.

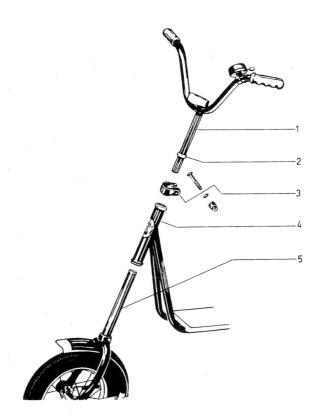

Location and Orientation

Of one thing we can be sure: we do not see things in the way common sense says we should.

Sir Francis Crick

"Place the hinge in the upper left corner." "Reverse and insert upside down over. . . ." "Place the batteries into. . . ." *Preposition time.*

You'd better hope the designer of your manual explained location and orientation right. Or d-i-y (do-it-yourself) becomes f-o-y (find-out-yourself). The result may be costly.

The Naumann Ideal Model D typewriter (Germany, 1927) came with a booklet that featured transparent overlays. Each overlay depicted a part of the machine that could be removed, and the pages were arranged in the proper order in which removal should be done.

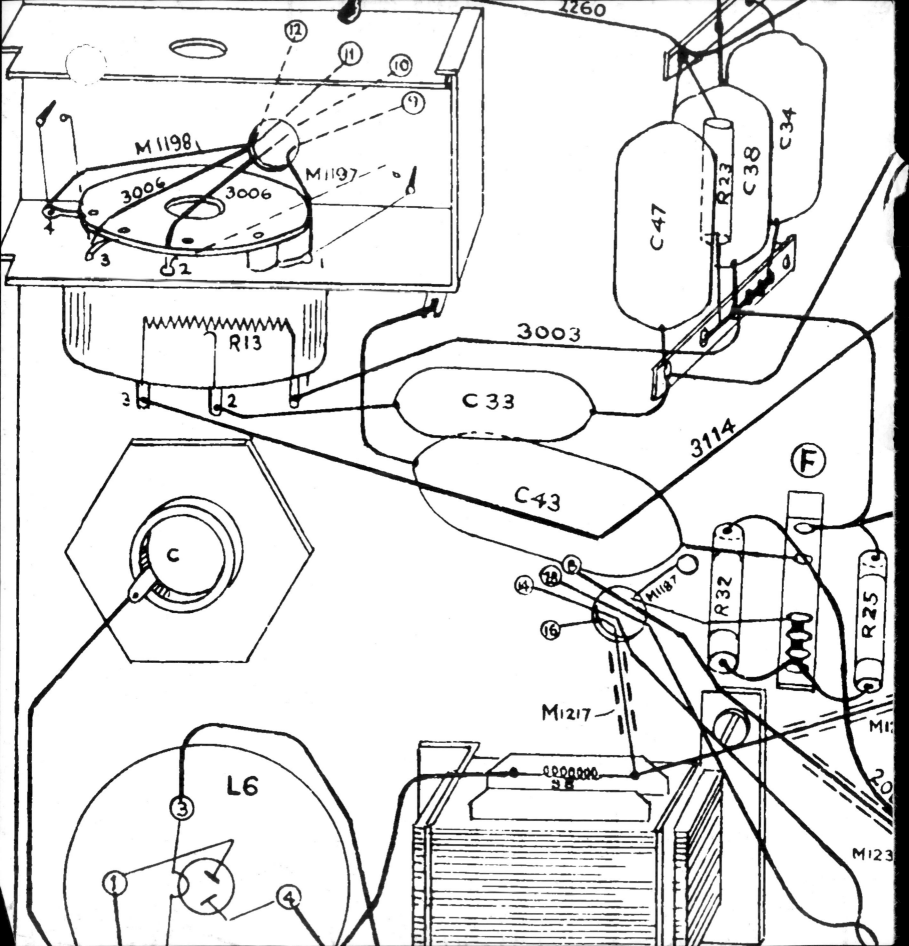

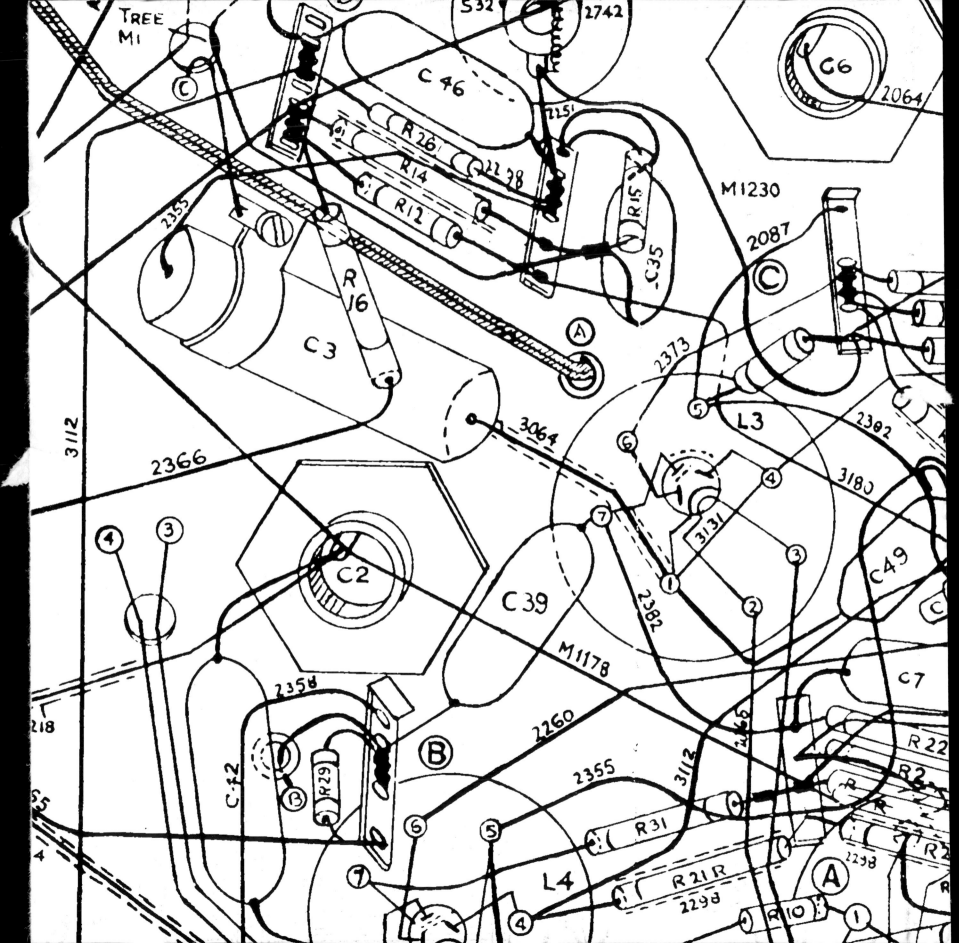

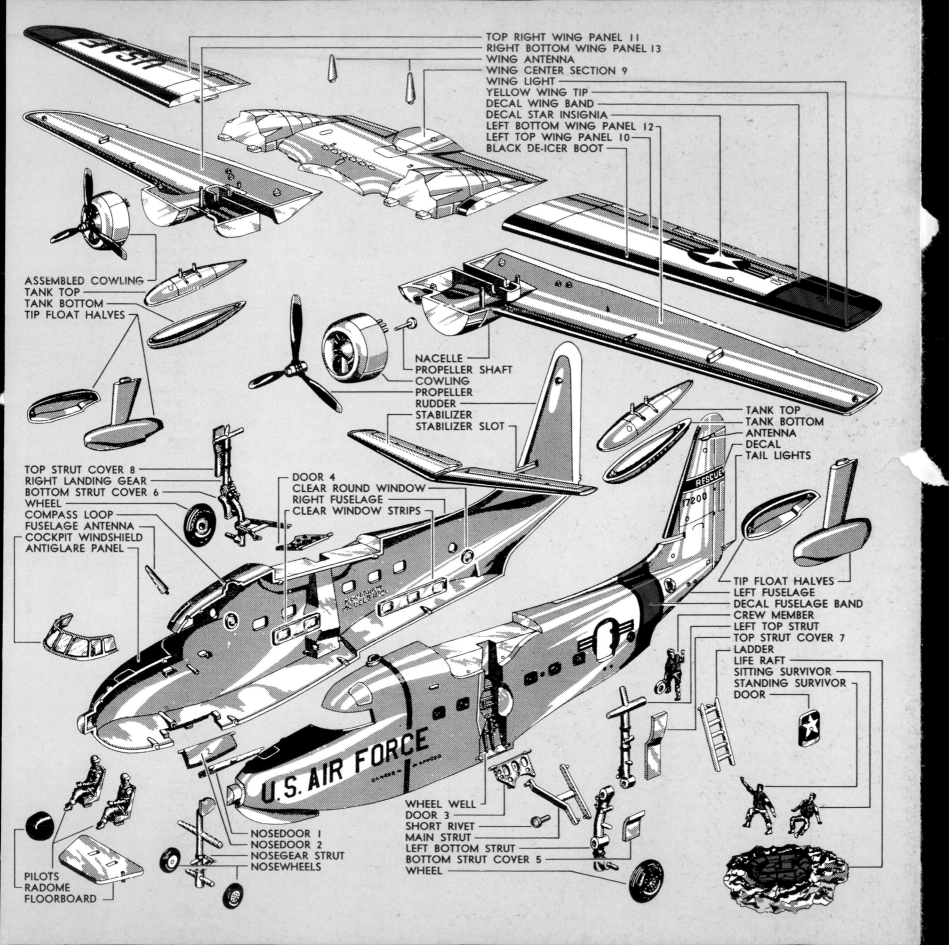

TOP RIGHT WING PANEL 11
RIGHT BOTTOM WING PANEL 13
WING ANTENNA
WING CENTER SECTION 9
WING LIGHT
YELLOW WING TIP
DECAL WING BAND
DECAL STAR INSIGNIA
LEFT BOTTOM WING PANEL 12
LEFT TOP WING PANEL 10
BLACK DE-ICER BOOT

ASSEMBLED COWLING
TANK TOP
TANK BOTTOM
TIP FLOAT HALVES

NACELLE
PROPELLER SHAFT
COWLING
PROPELLER
RUDDER
STABILIZER
STABILIZER SLOT

TANK TOP
TANK BOTTOM
ANTENNA
DECAL
TAIL LIGHTS

RESCUE

TOP STRUT COVER 8
RIGHT LANDING GEAR
BOTTOM STRUT COVER 6
WHEEL
COMPASS LOOP
FUSELAGE ANTENNA
COCKPIT WINDSHIELD
ANTIGLARE PANEL

DOOR 4
CLEAR ROUND WINDOW
RIGHT FUSELAGE
CLEAR WINDOW STRIPS

TIP FLOAT HALVES
LEFT FUSELAGE
DECAL FUSELAGE BAND
CREW MEMBER
LEFT TOP STRUT
TOP STRUT COVER 7
LADDER
LIFE RAFT
SITTING SURVIVOR
STANDING SURVIVOR
DOOR

U.S. AIR FORCE

PILOTS
RADOME
FLOORBOARD

NOSEDOOR 1
NOSEDOOR 2
NOSEGEAR STRUT
NOSEWHEELS

WHEEL WELL
DOOR 3
SHORT RIVET
MAIN STRUT
LEFT BOTTOM STRUT
BOTTOM STRUT COVER 5
WHEEL

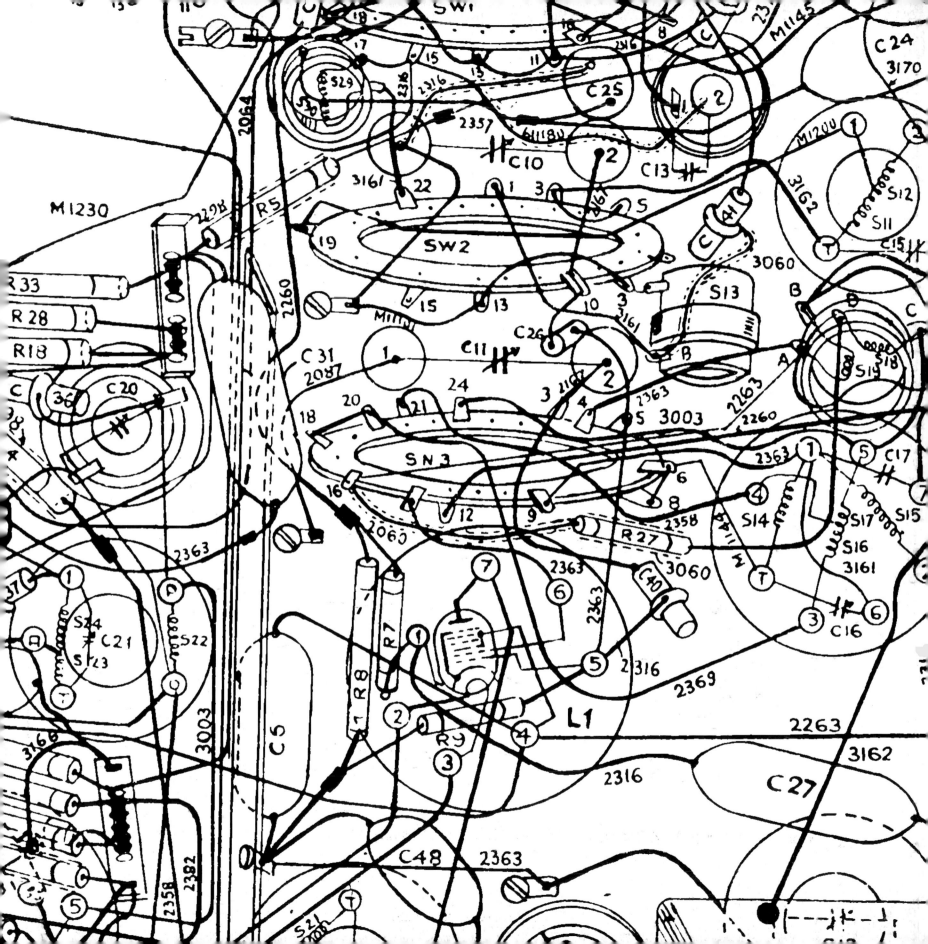

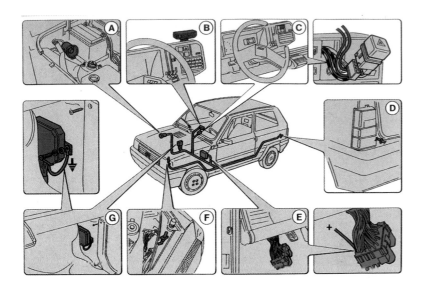

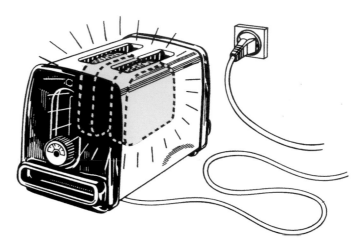

The location and orientation of elements are sometimes indicated using ghost views. Call-outs may indicate details.

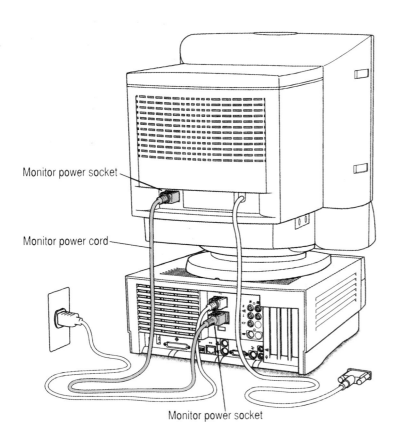

Monitor power socket

Monitor power cord

Monitor power socket

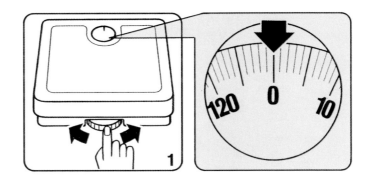

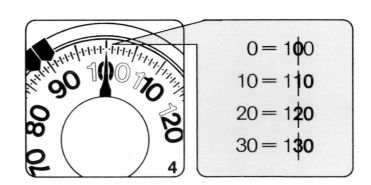

0 = 100
10 = 110
20 = 120
30 = 130

The Art of Instructional Design

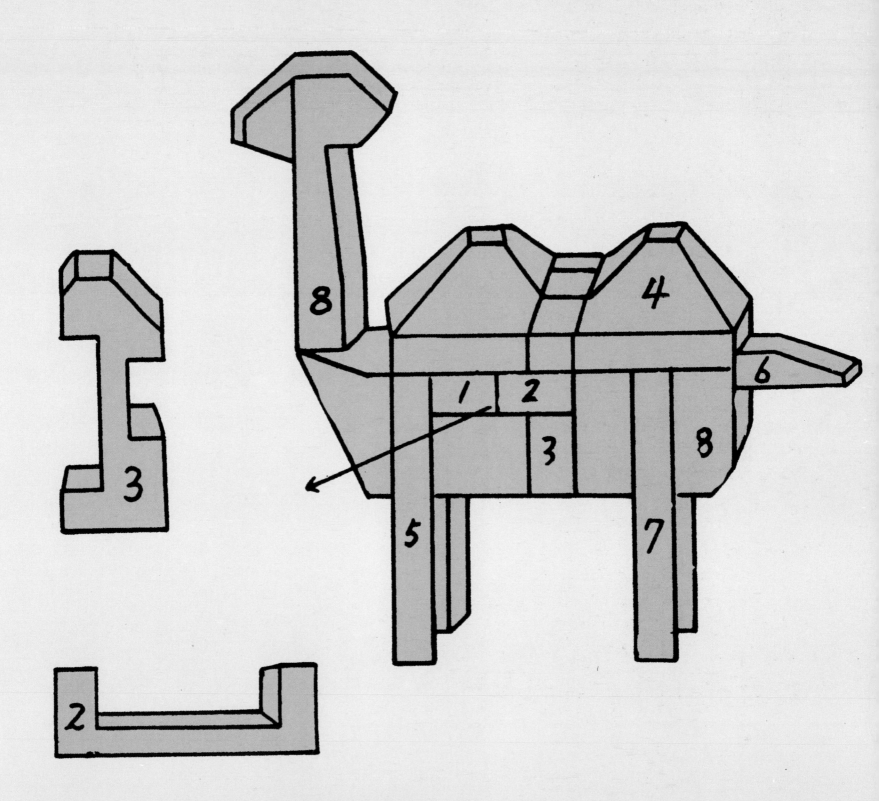

The location of an element may also be indicated with a color highlight or a frame, but most often the hand of the master points the way. Here, it's a secretarial fairy, at her Gabriëlle typewriter, who shows us which knob to turn.

The Art of Instructional Design

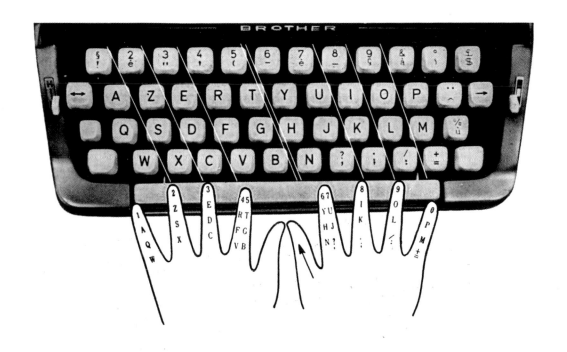

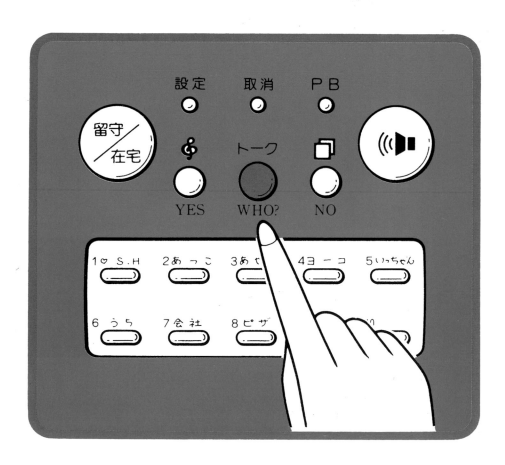

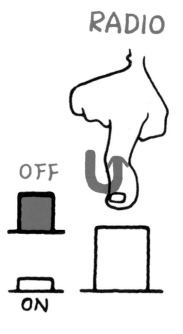

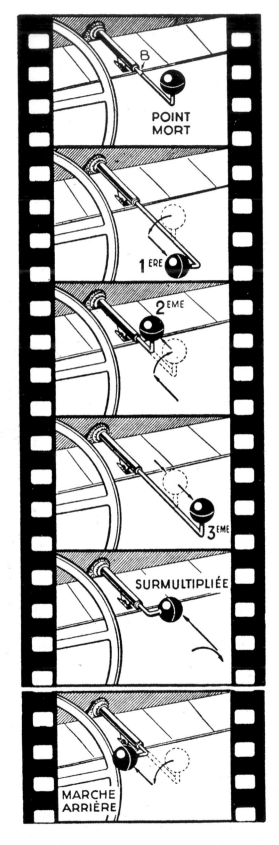

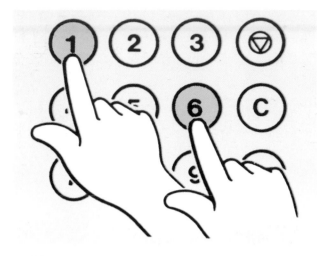

Sequences

Am Heimcomputer sitz' ich hier
Und programmier die Zukunft mir.
[I've programmed my home computer
And beamed myself into the future]
Kraftwerk, *Computerwelt* (1981)

1. Hold the end . . . 2. Move the slider . . .
3. Secure the end . . . 4. Move the slider
back . . . 5. Pull the strap. . . . *The ordinals and*
cardinals.

First do this, then do that, next . . . until Step
17c. It seems easy to indicate the order of
working: 1, 2, 3 or A, B, C or maybe I, II, III for
the Latinists among us. From the top down or
left to right? In columns or per page? A
flowchart might help, or a cartoon-type series
of illustrations. A row of clocks next to a row
of pictures. . . . But let's hope the designer
isn't too creative in this point.

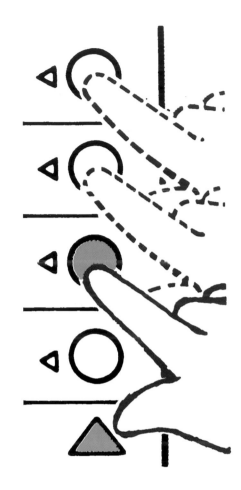

Sequences are often shown
movie-style, in series of
pictures, sometimes even in
film frames. Which scenes
are you missing in these
movies?

The Art of Instructional Design

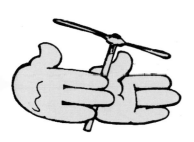

1

2

3

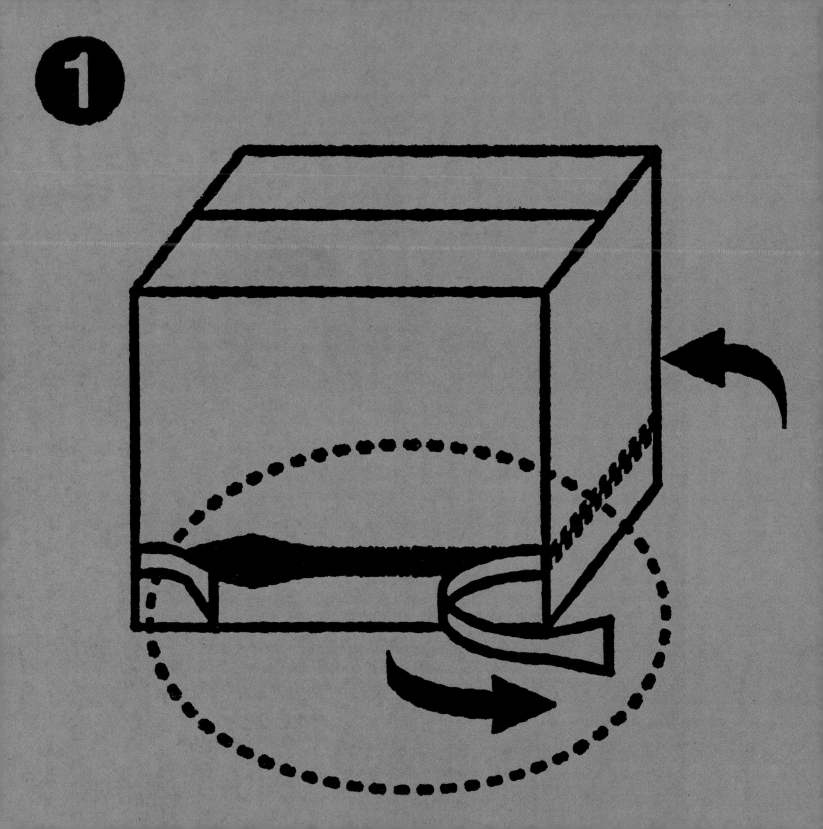

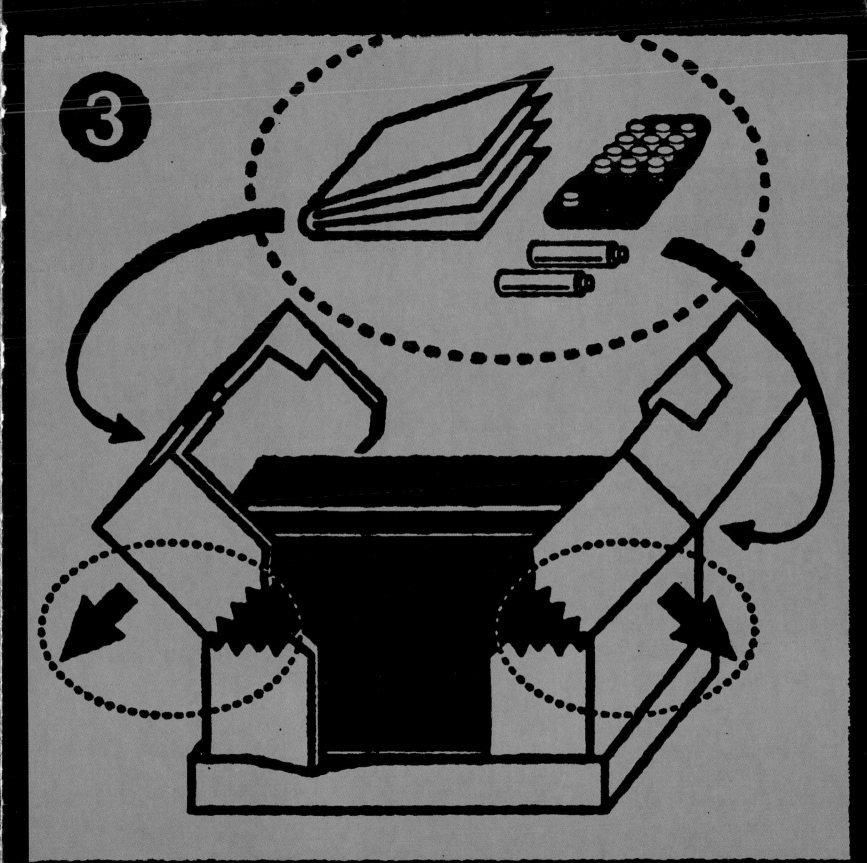

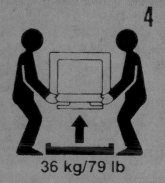

1 **2** **3** **4**

36 kg/79 lb

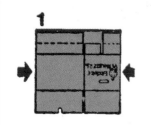
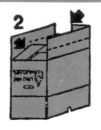

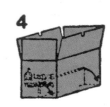

1 **2** **3** **4** **5**

UNPACKING PROCEDURE

UNPACKING

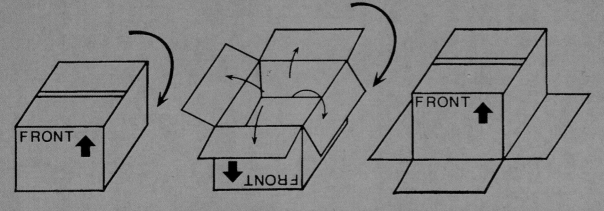

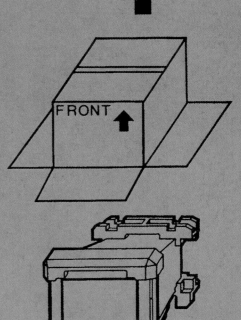

| Turn the carton upside-down | Open the bottom flaps | Turn the carton upside-down again | Pull the carton up carefully |

RE-PACKING

TEAR HERE
↑ **↑**
PUSH BACK   PUSH BACK
TO OPEN
PUSH UP WINGS
AND PRESS BACK
TO FORM POURING SPOUT

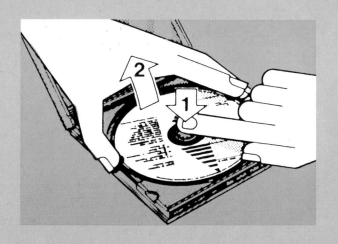

The Art of Instructional Design

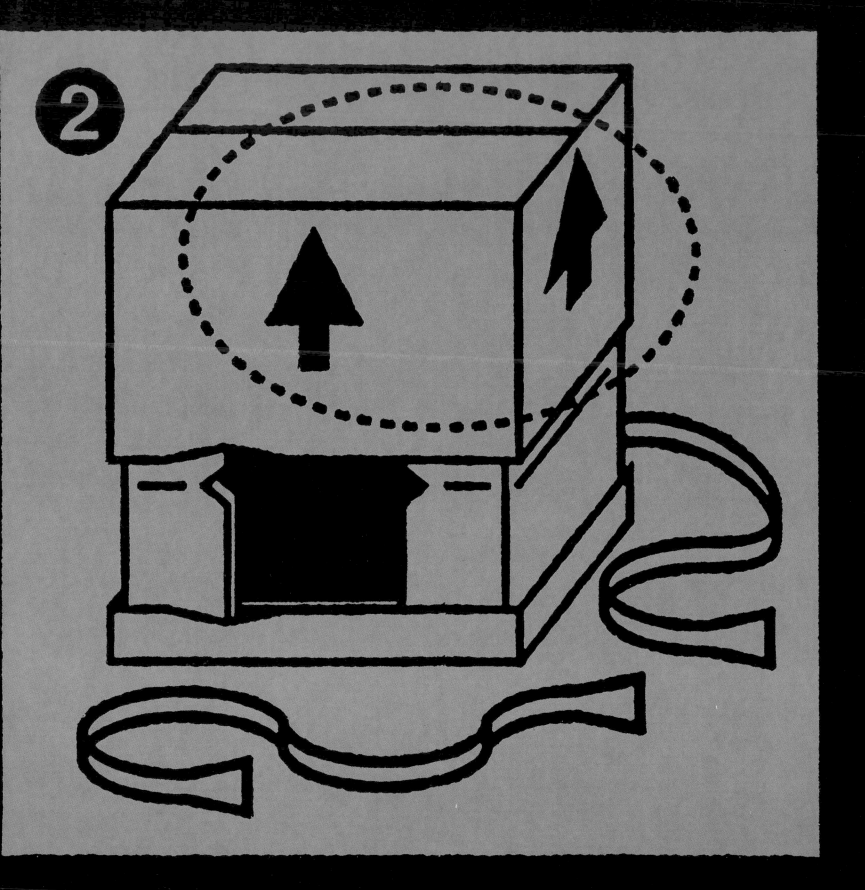

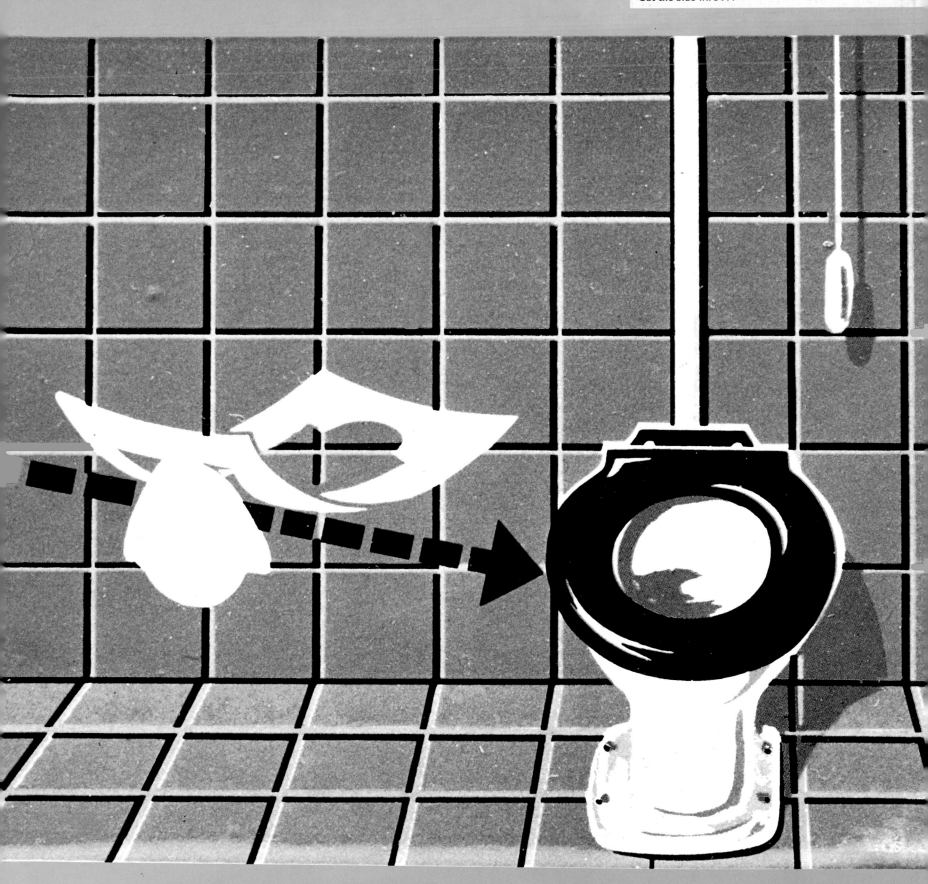

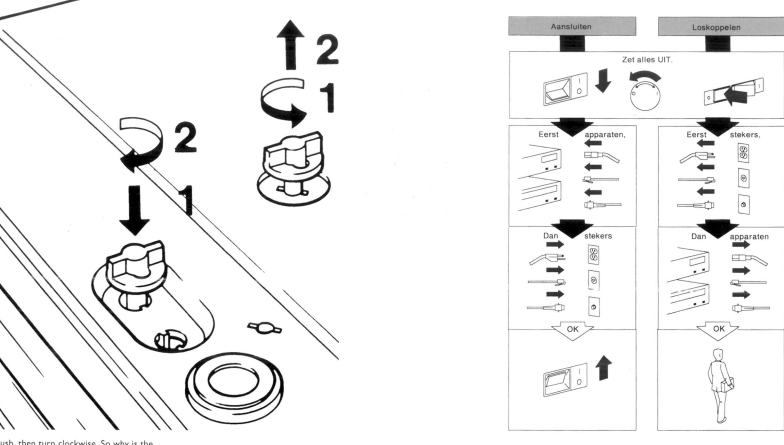

First push, then turn clockwise. So why is the
"2" on top (which means we read it first) and
the "1" underneath (which we read last)? And
vice versa: first turn counterclockwise, then
pull. Read from bottom to top again.

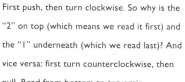

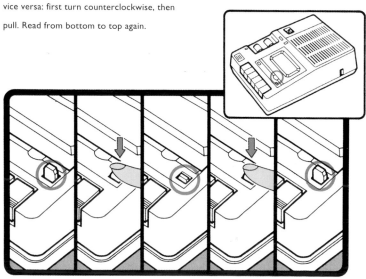

First A, then B, or simultaneously?

The Art of Instructional Design

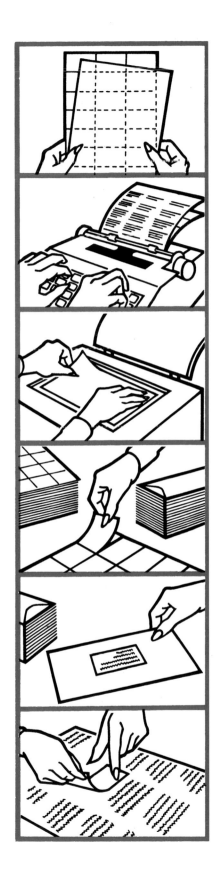

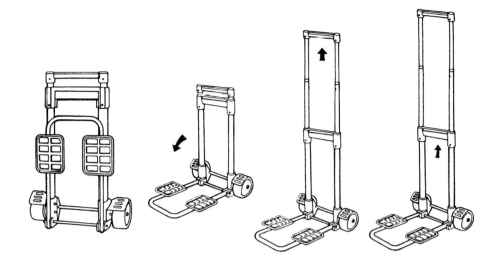

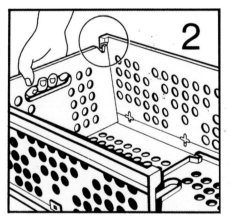

External calls

Someone is dialling in from outside the company.

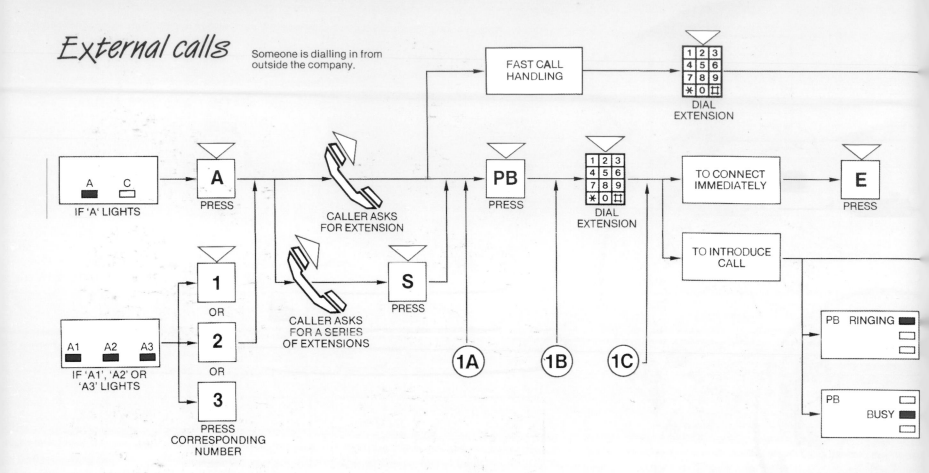

IF 'A' LIGHTS

A PRESS

CALLER ASKS FOR EXTENSION

IF 'A1', 'A2' OR 'A3' LIGHTS

1 OR **2** OR **3**
PRESS CORRESPONDING NUMBER

CALLER ASKS FOR A SERIES OF EXTENSIONS

S PRESS

FAST CALL HANDLING

DIAL EXTENSION

PB PRESS

DIAL EXTENSION

TO CONNECT IMMEDIATELY

E PRESS

TO INTRODUCE CALL

(1A)   (1B)   (1C)

PB RINGING

PB BUSY

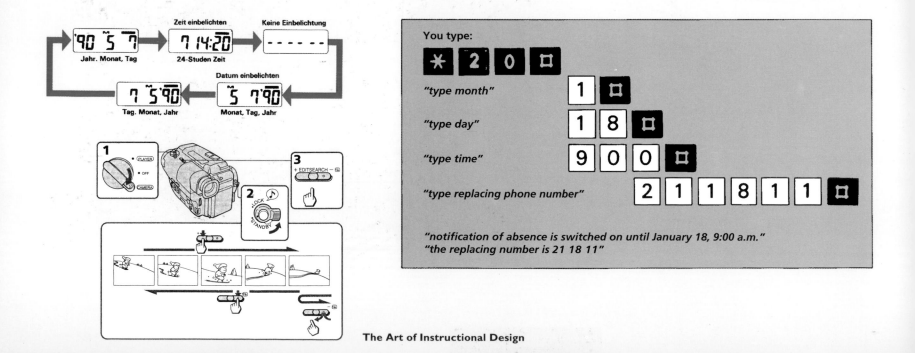

Zeit einbelichten   Keine Einbelichtung

'90 5 7
Jahr. Monat, Tag

7 14:20
24-Studen Zeit

- - - - -

Datum einbelichten

7 5'90
Tag. Monat, Jahr

5 7'90
Monat, Tag, Jahr

1 PLAYER • OFF CAMERA

2 LOCK ♪ STANDBY

3 + EDITSEARCH – 🖑

You type:

✳ **2** 0 ⊞

"type month"   **1** ⊞

"type day"   **1 8** ⊞

"type time"   **9 0 0** ⊞

"type replacing phone number"   **2 1 1 8 1 1** ⊞

"notification of absence is switched on until January 18, 9:00 a.m."
"the replacing number is 21 18 11"

The Art of Instructional Design

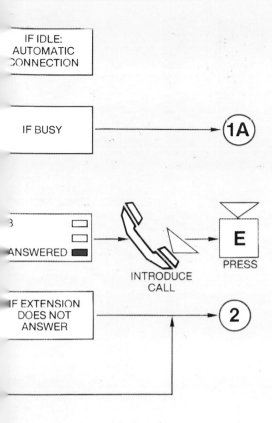

A technical flowchart for a telephone operator—assumed to be a woman in this case? (Judging from the font type used for "External calls.")

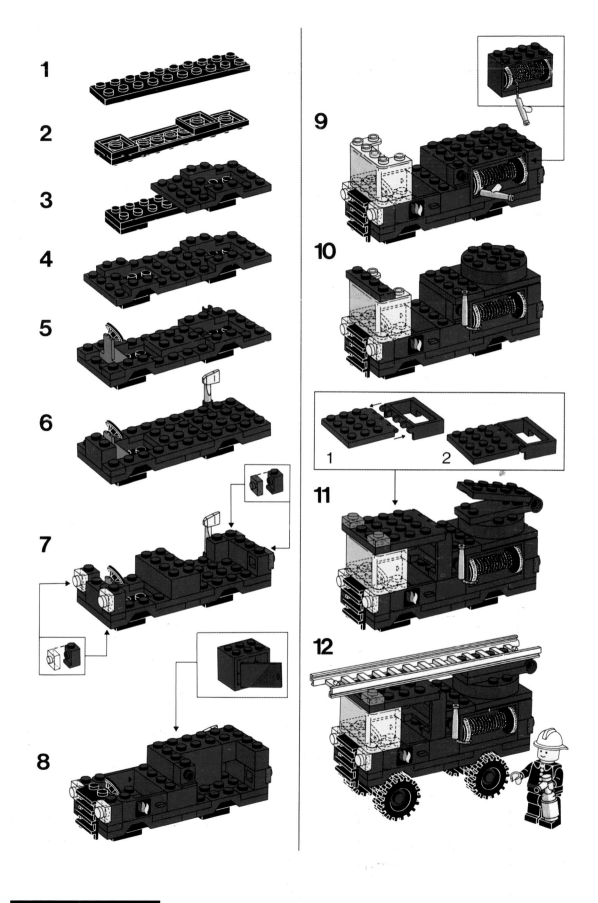

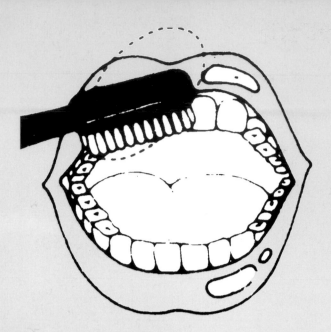

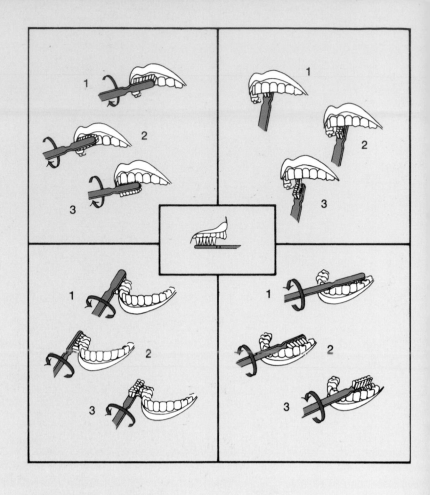

Movements

Ease of use and ease of learning are not the same.

Katherine Haramundanis, *The Art of Technical Documentation* (1992)

"Rotate," "pull," "shake," "press," "hold," "lift," "push," "flip," "fit," "bend," "twist," "insert," "unscrew," "fold," "turn counterclockwise." *The verbs explaining how to do it.*

Combinations of movements produce the most impressive arrows: in 3-D, or with twisted lines, graduated shading or color, or numbers. "Press and turn," "Lift, then press"— complex actions require Pop Art arrows. Dotted lines indicate stages of a movement—we may have learned how to read these from cartoons.

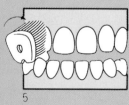
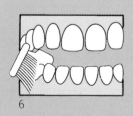
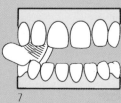
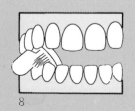

2   3   4   5   6   7   8

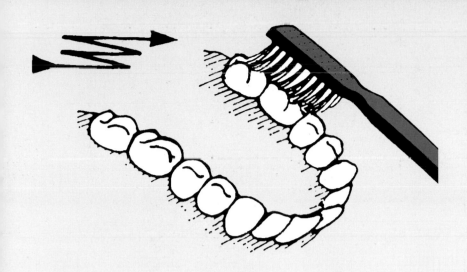

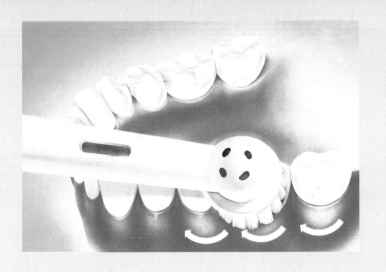

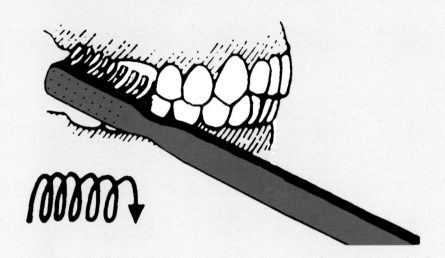

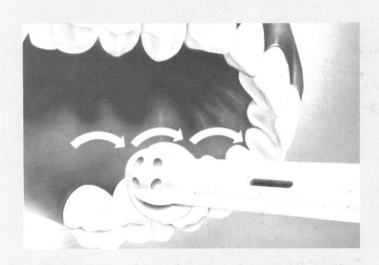

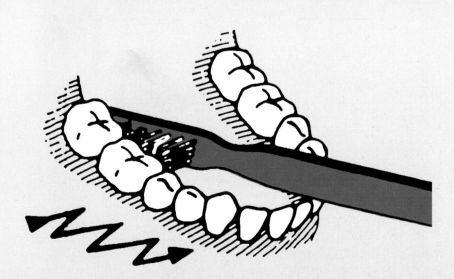

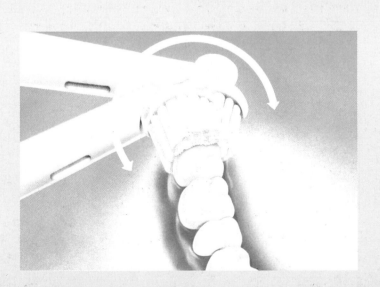

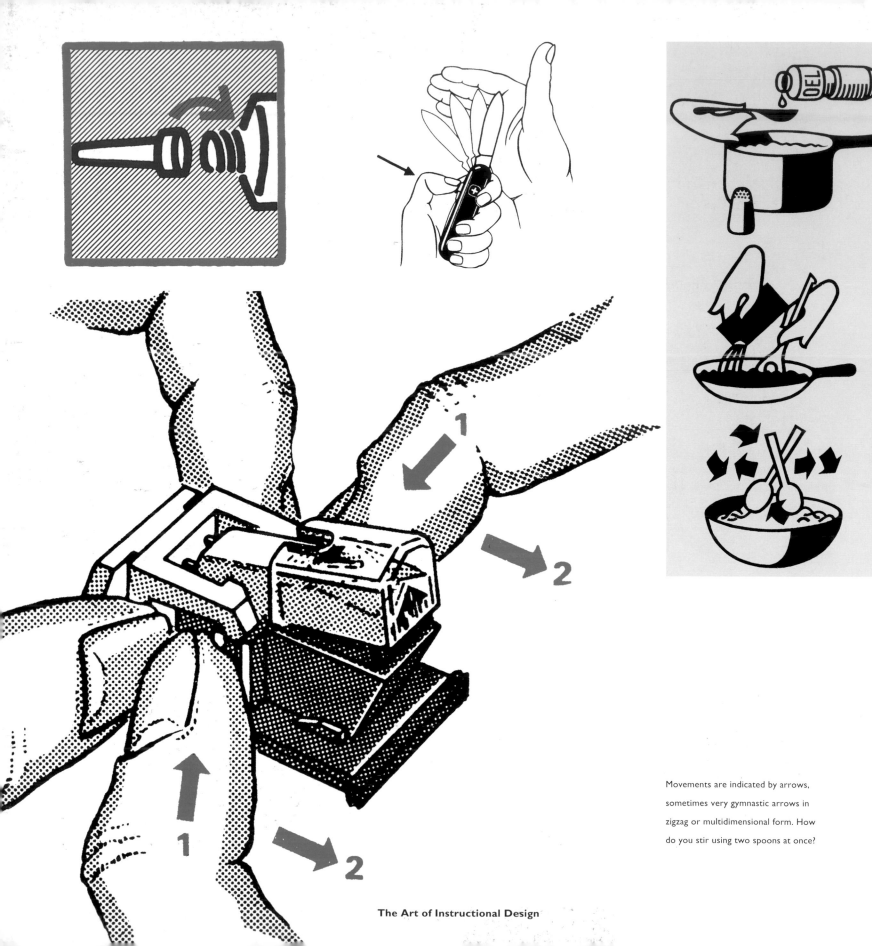

Movements are indicated by arrows, sometimes very gymnastic arrows in zigzag or multidimensional form. How do you stir using two spoons at once?

The Art of Instructional Design

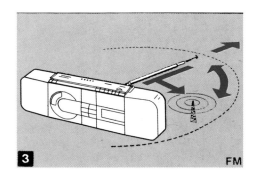

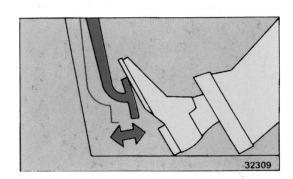

APRIRE QUI ↑

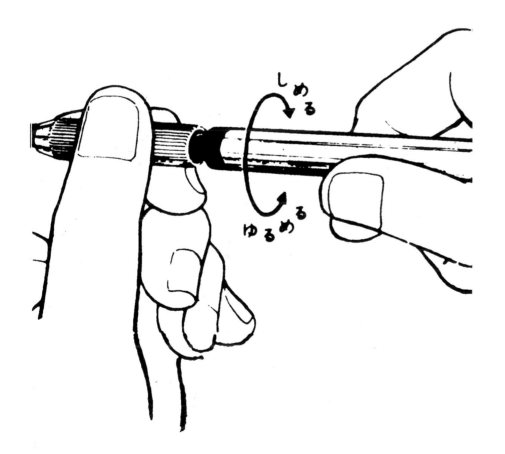

しめる

ゆるめる

PRESS →

The Art of Instructional Design

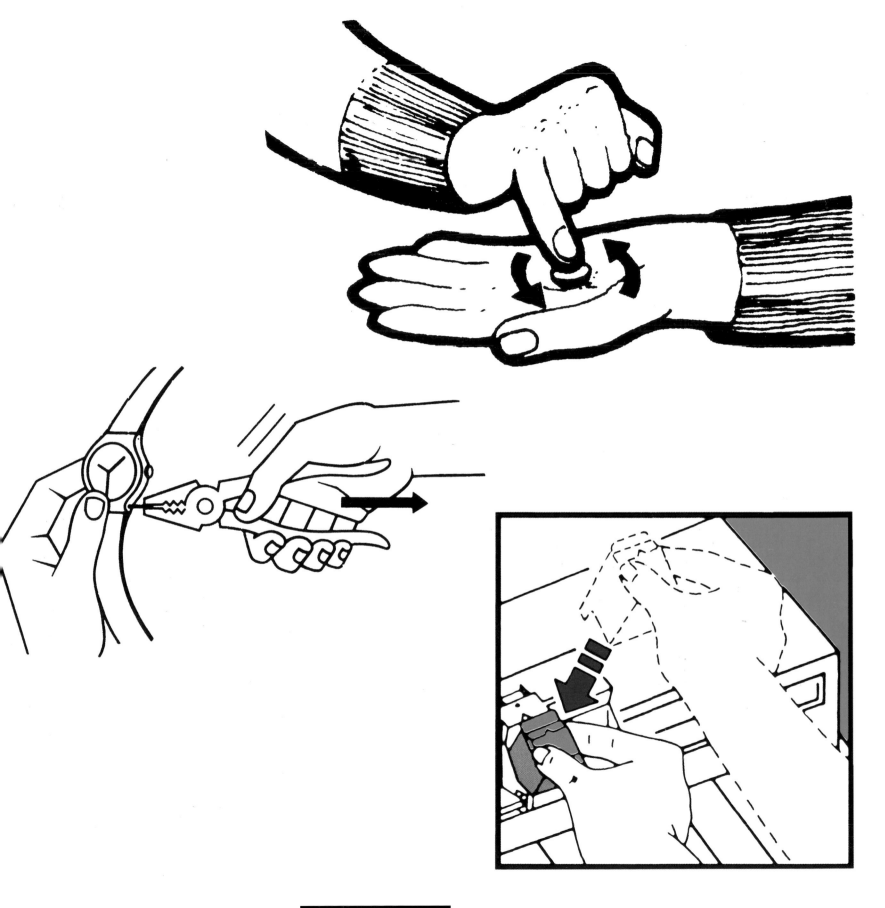

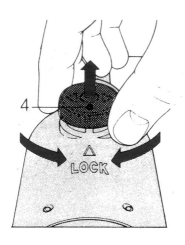

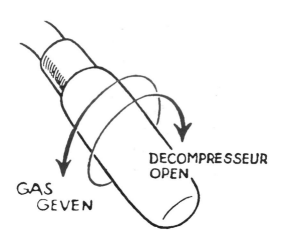

DECOMPRESSEUR
OPEN

GAS
GEVEN

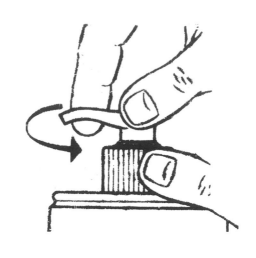

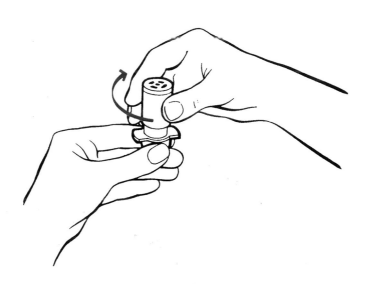

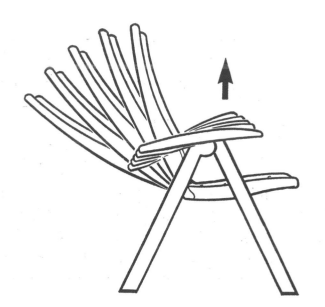

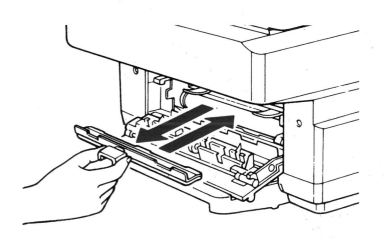

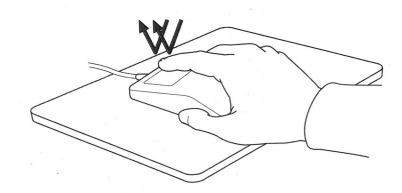

The Art of Instructional Design

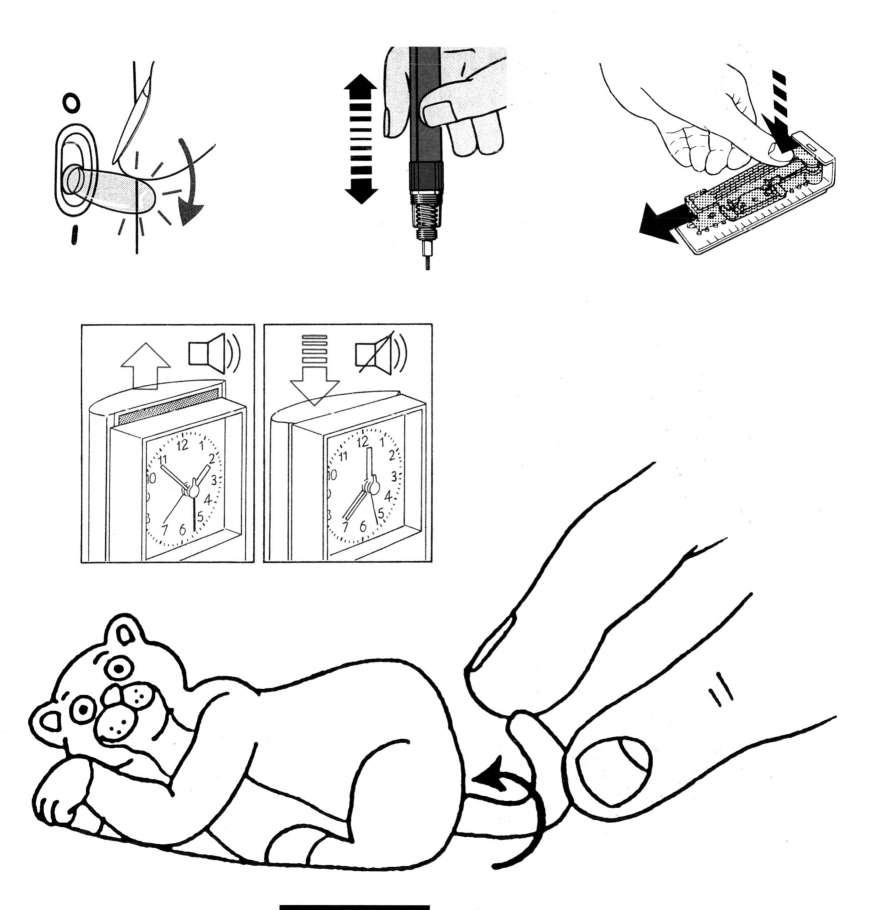

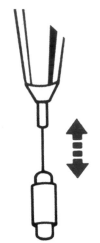

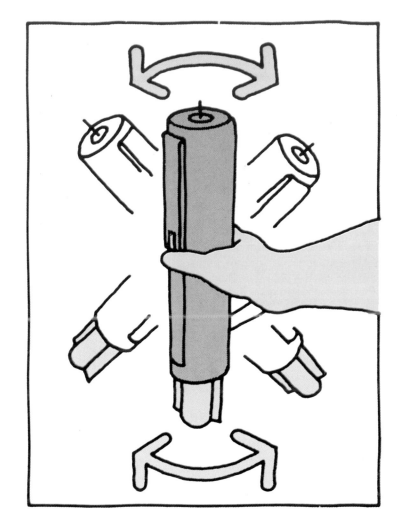

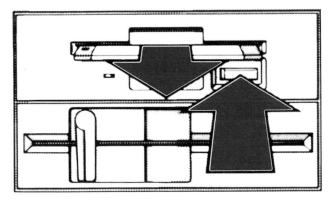

Shaking may be indicated by illustrating several different positions in one frame, like a triple- or quadruple-exposed photograph.

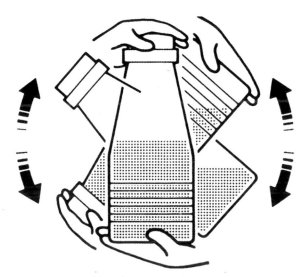

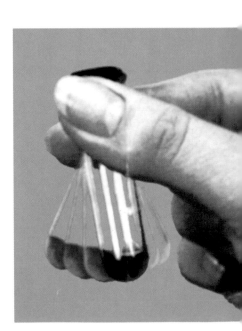

The Art of Instructional Design

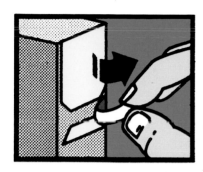

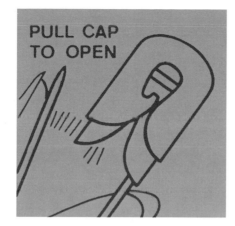

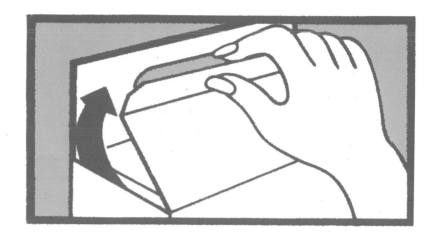

PULL CAP
TO OPEN

PUSH CAP
TO LOCK

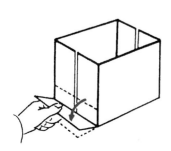

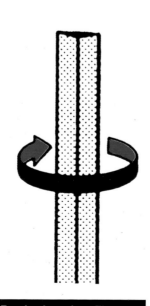

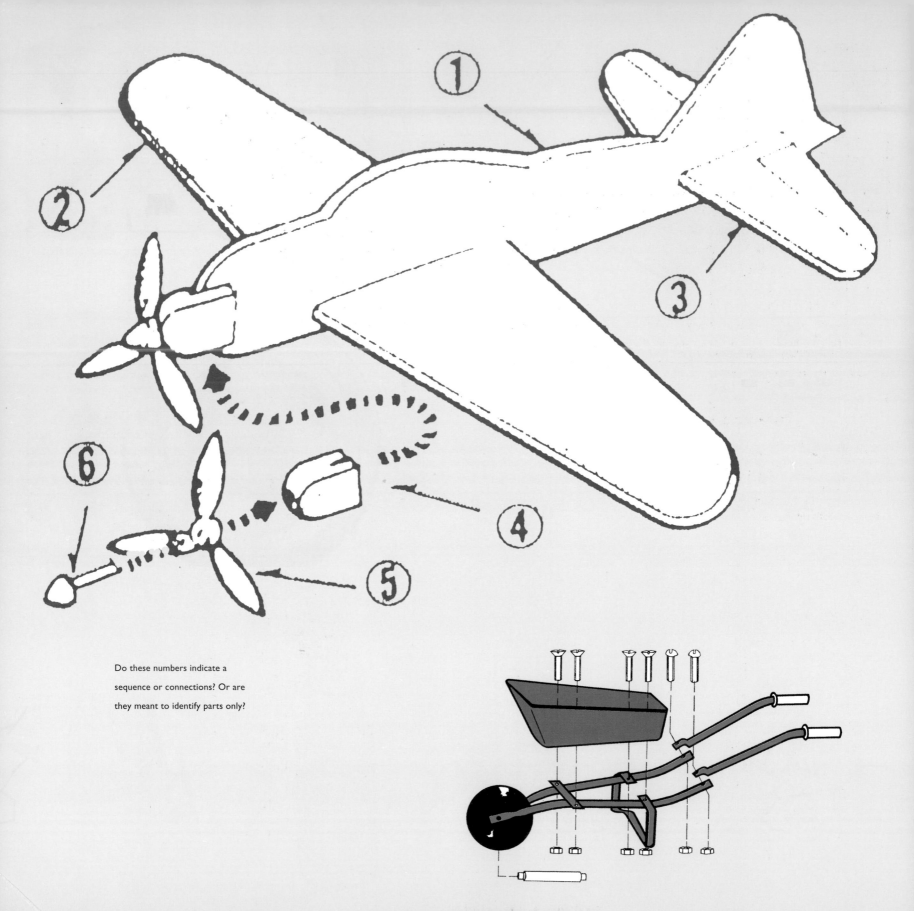

Do these numbers indicate a
sequence or connections? Or are
they meant to identify parts only?

Connections

Erst in der Beschränkung zeigt sich der Meister.
[Conciseness shows masterhood.]

Johann Wolfgang von Goethe

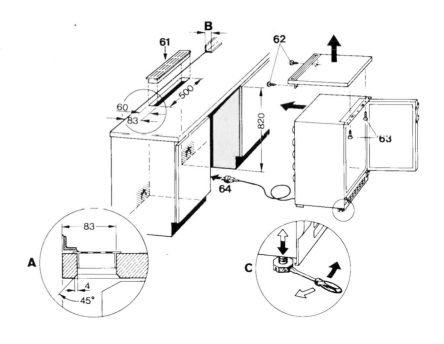

"Be sure the braided shield touches the clamp." "Insert the DC cable on the connecting plate into. . . ." "Open the cover by sliding it downward and then swinging it outward." "Connect cable 13 to socket 4." *Prepositions and verbs of contact.*

Magnified inserts and frames show us what to connect to what. We set up our nearly professional-quality stereo with the help of a semi-professional electronic schema showing where to connect which plug. And arrows, of course, always arrows.

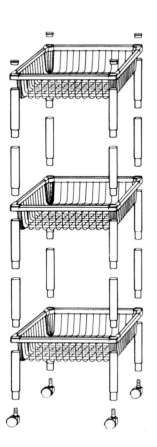

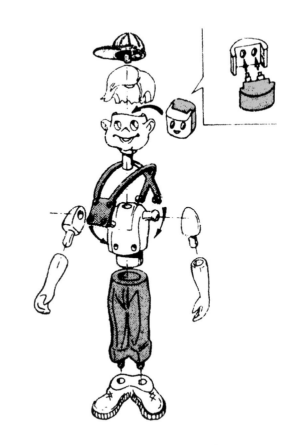

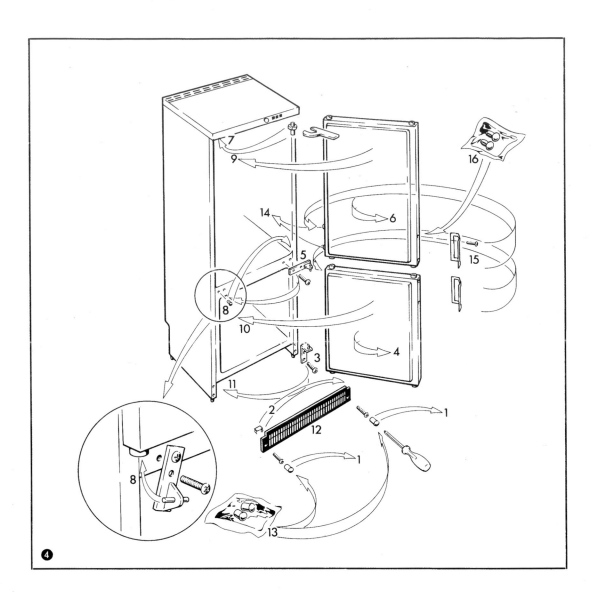

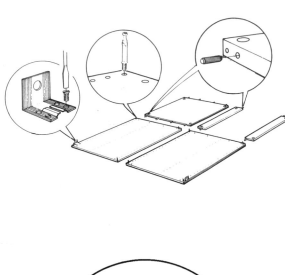

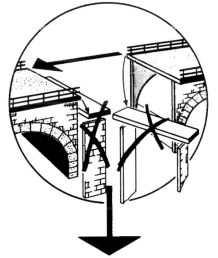

"If there is a way to do it wrong, he'll find it."

The original formulation of Murphy's Law. In fact, this was first said about Ed Murphy, an aeronautical engineer from California.

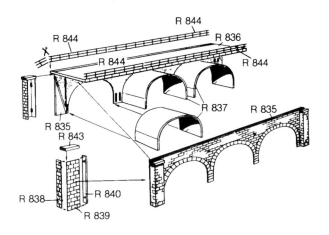

The Art of Instructional Design

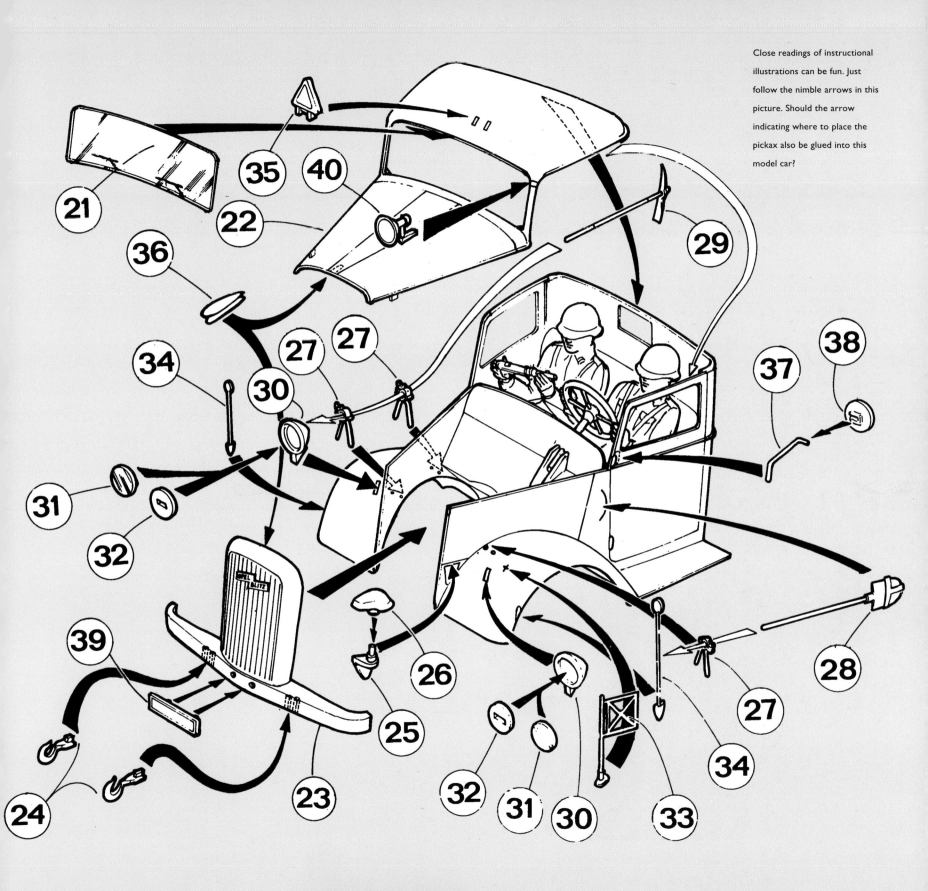

Close readings of instructional illustrations can be fun. Just follow the nimble arrows in this picture. Should the arrow indicating where to place the pickax also be glued into this model car?

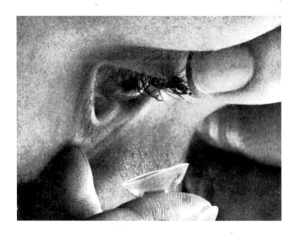

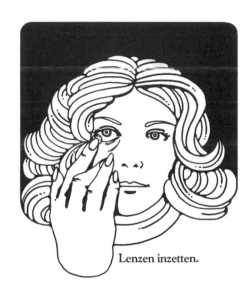

Lenzen inzetten.

Vloeistof
weggooien.

Action!

Dictum factum
[Done as said]
Terence, *Heautontimoroumenos* (904)

"Check," "Adjust," "Use," "Take out." *Verbs again.*

A metal saw telling us to cut here. Scissors showing where to clip the coupon out of an advertisement. On/off switches in the "on" position—or in the "off" position plus an arrow—or two pictures of the switch: one in the "off" position and the next in the "on" position—or the same thing plus an arrow for extra clarity—or too much clarity. "Personal" demonstrators—illustrated teachers with the very best of intentions.

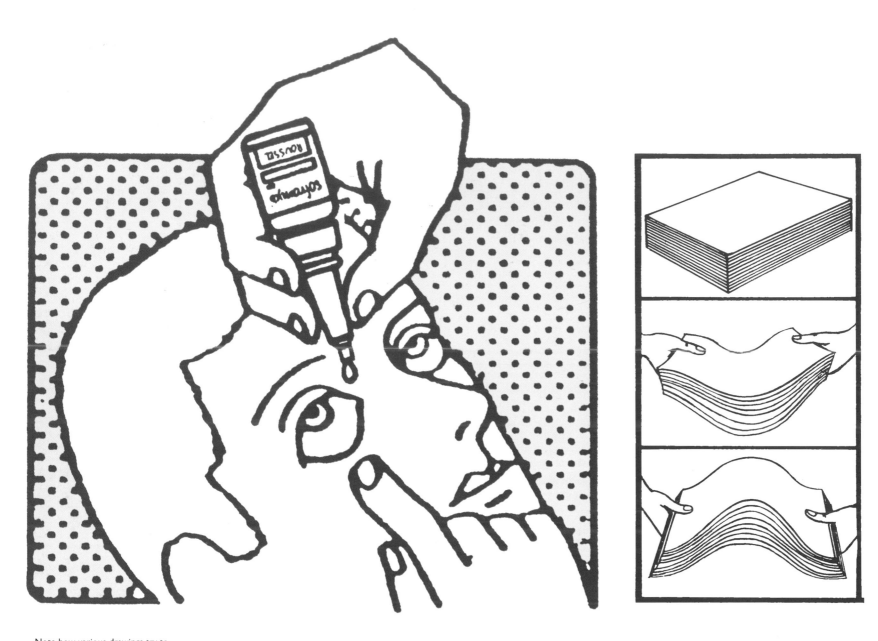

Note how various drawings try to coerce you into action by taking different approaches: sympathetic, harsh, friendly, respectful, cynical, childish, exaggerated, dreamy, tempting and everything in-between.

The Art of Instructional Design

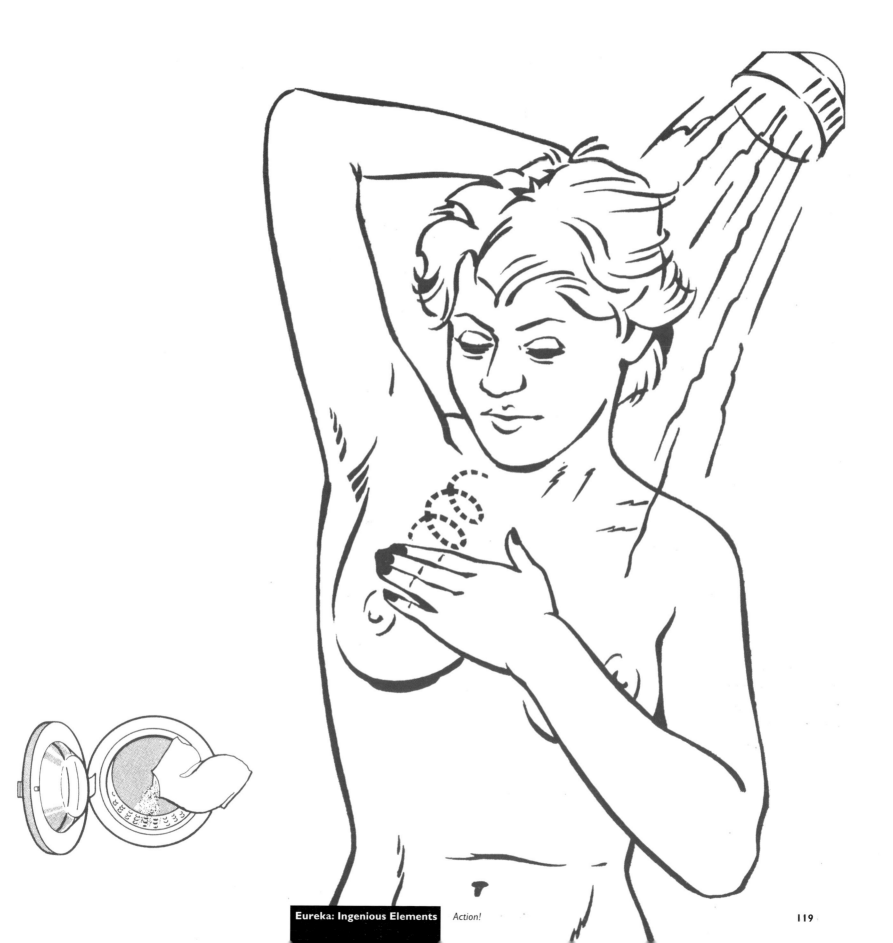

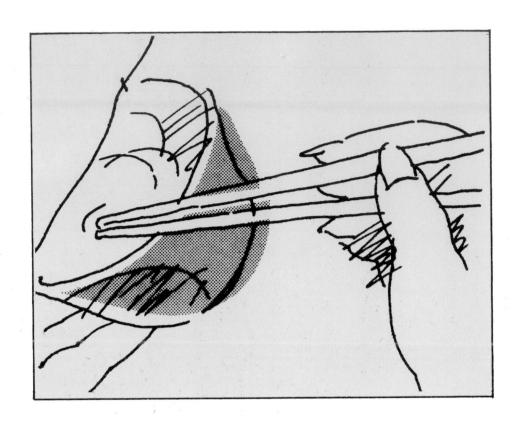

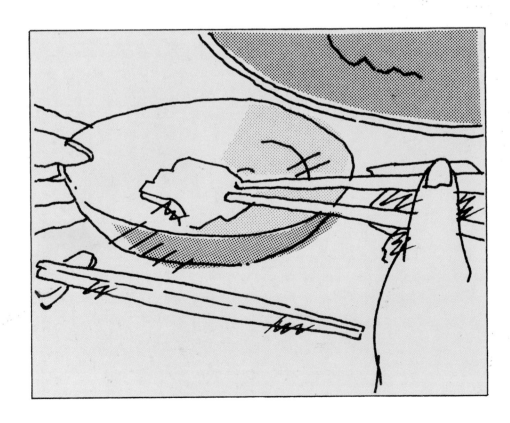

The Art of Instructional Design

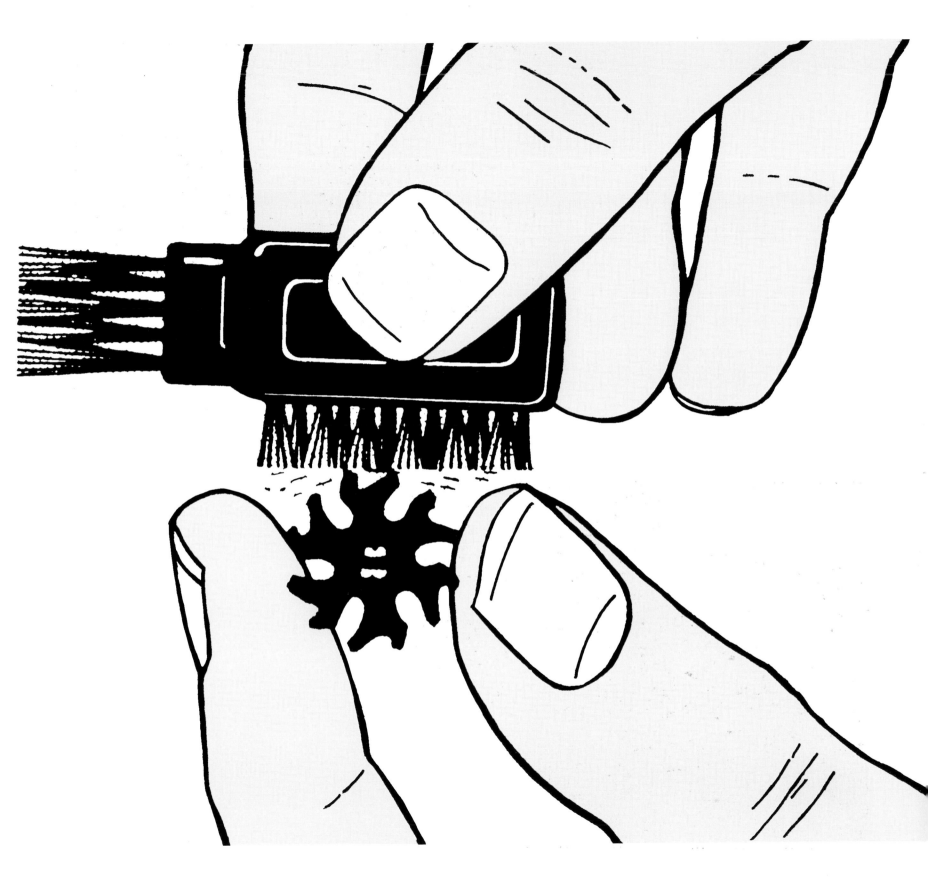

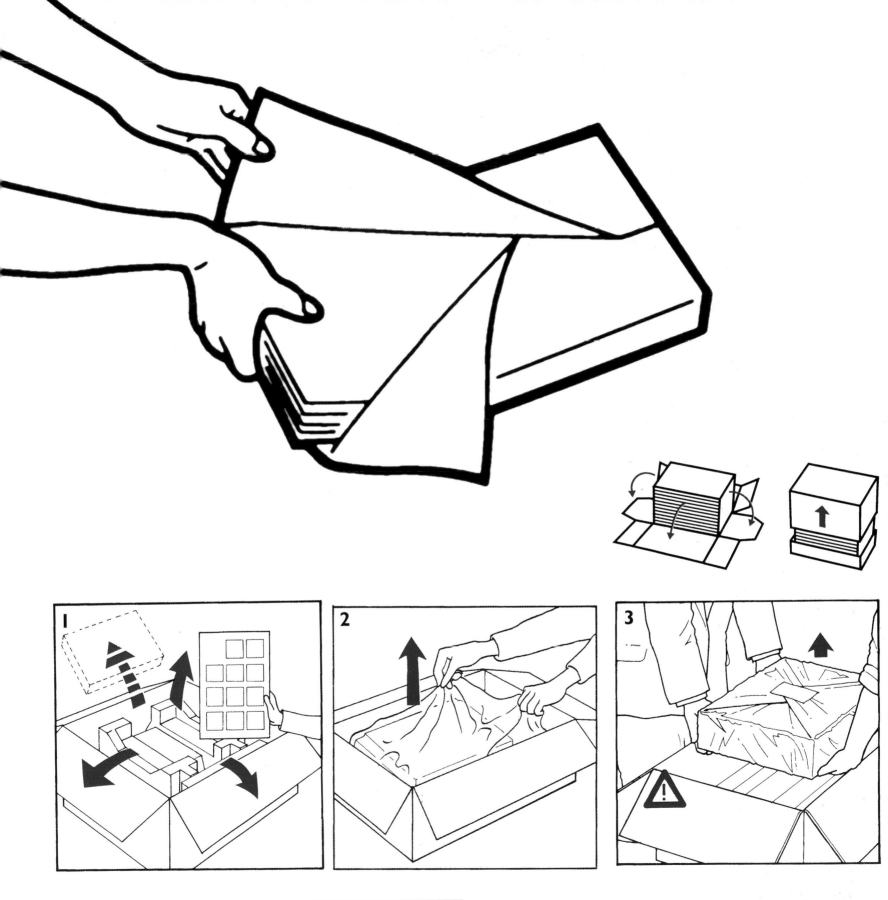

  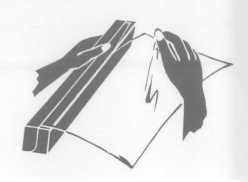

The Art of Instructional Design

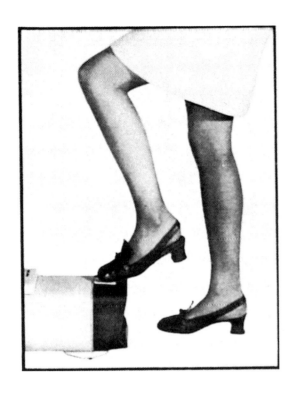

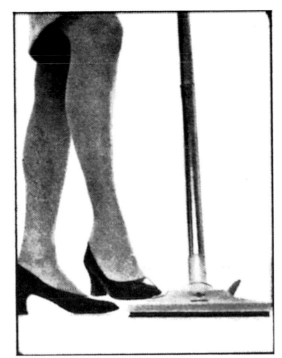

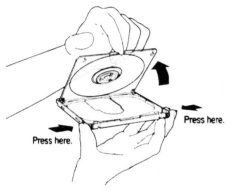

Press here.

Press here.

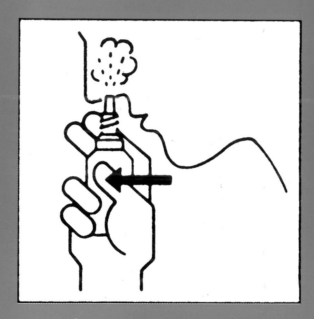

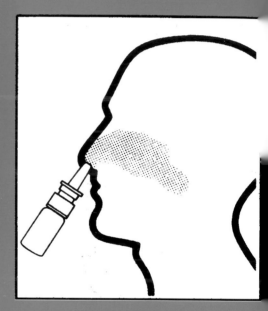

Five ways to use a nasal spray;

five dialects of graphic language.

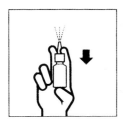

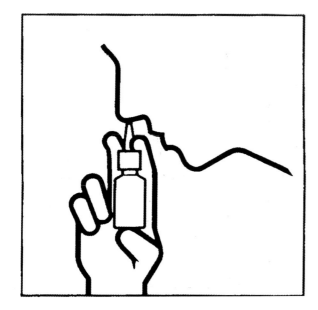

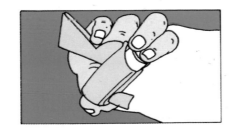

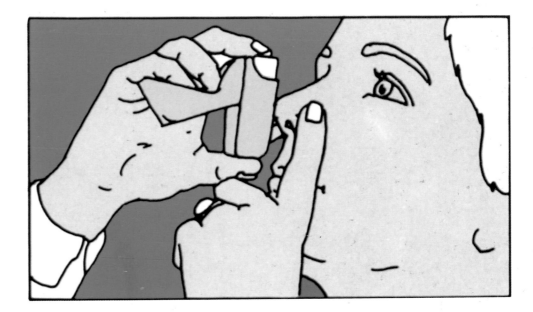

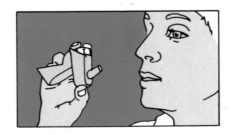

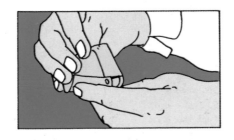

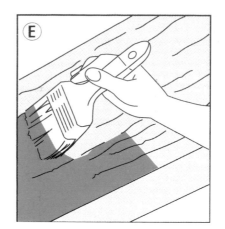

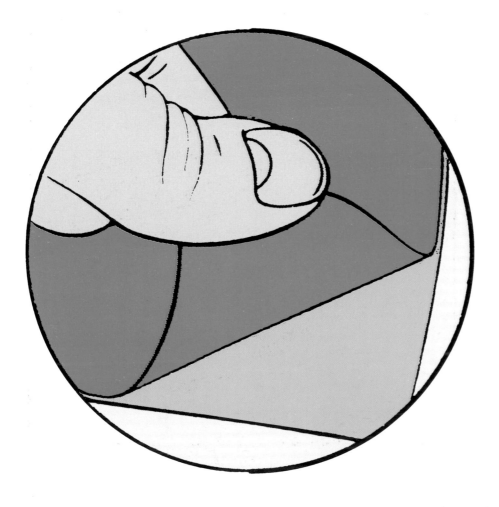

The Art of Instructional Design

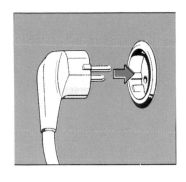

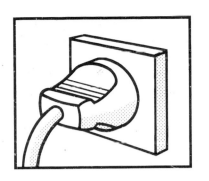
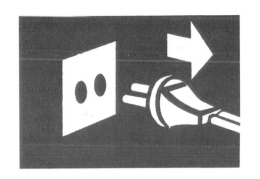

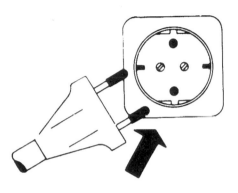
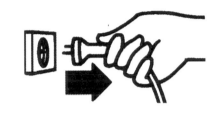
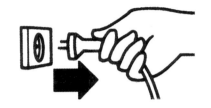

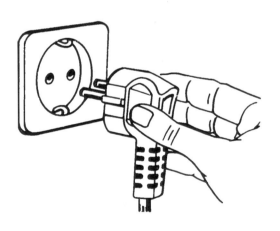

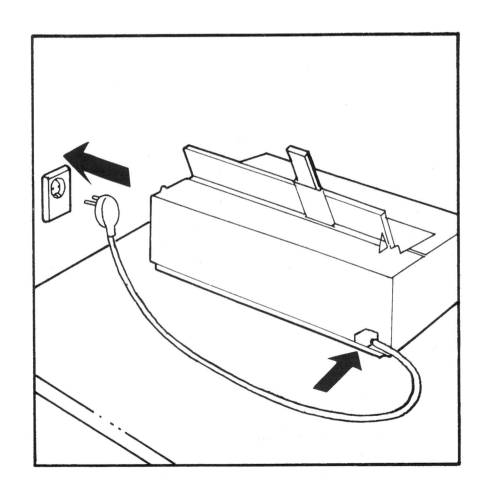

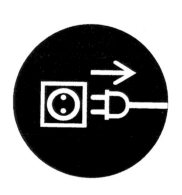

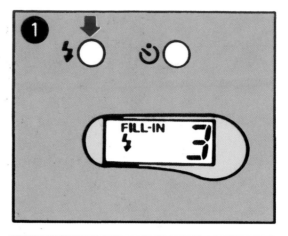

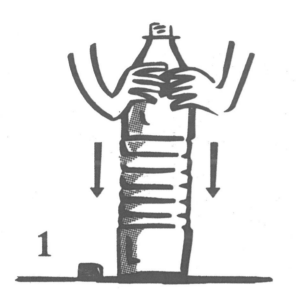

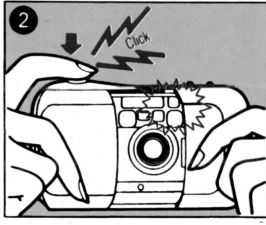

Cause and Effect

If all else fails, read the manual.

"Press MENU. The menu will disappear from the monitor." "Press XX; you will hear a signal." "When you press ● the remaining time will be displayed." "Press until you hear 'Click.'" *The visual conjunctions.*

Action = reaction. You press a button and the green light flashes on—at least it did in the manual.

The cartoon "Click" is making a career of it, just like the musical note. Designers often use arrows here to indicate *if > then*, as in logic notation.

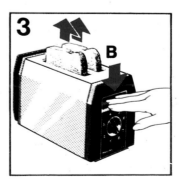

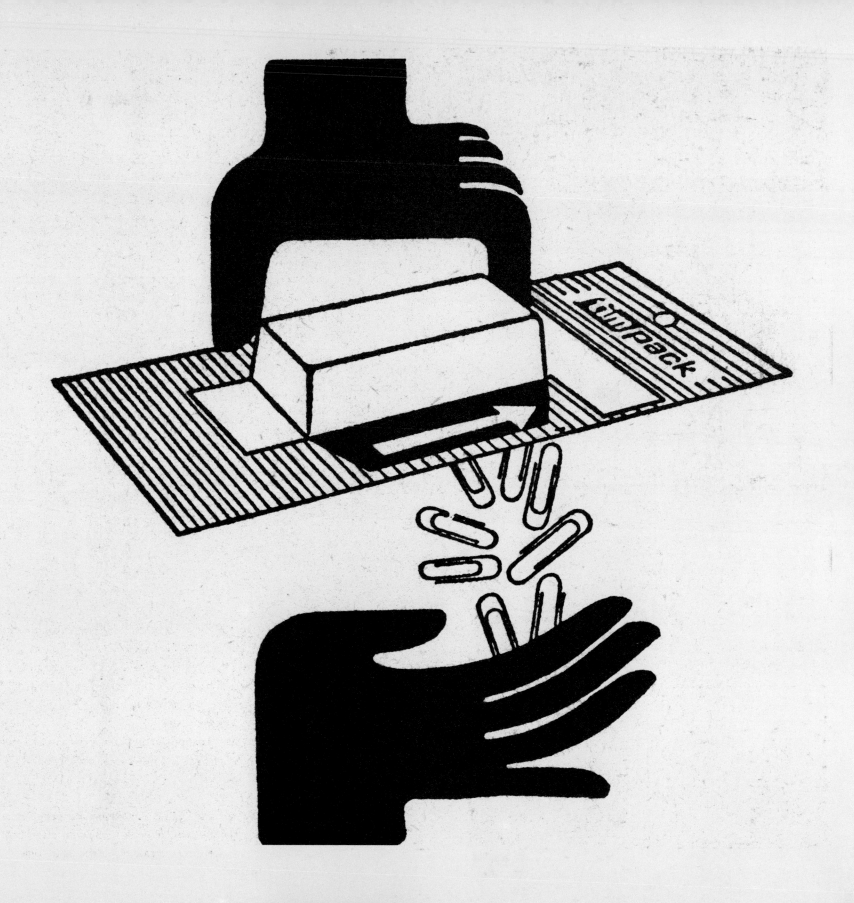

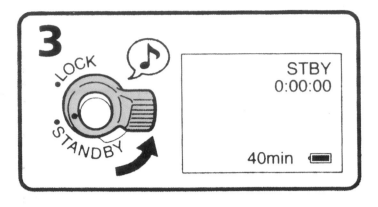

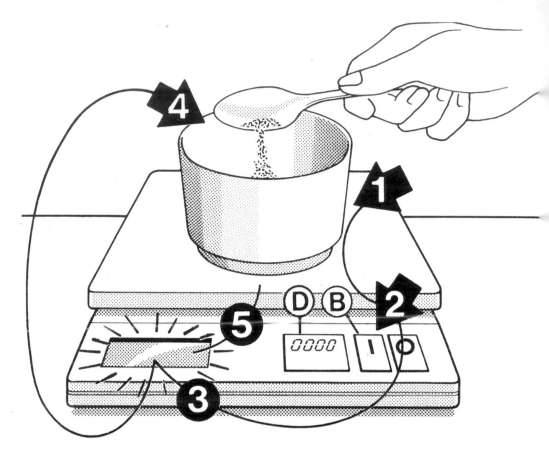

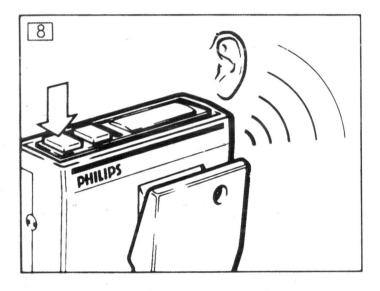

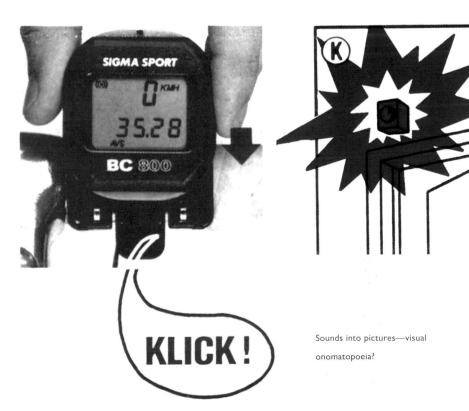

Sounds into pictures—visual onomatopoeia?

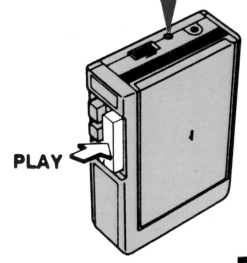

PLAY

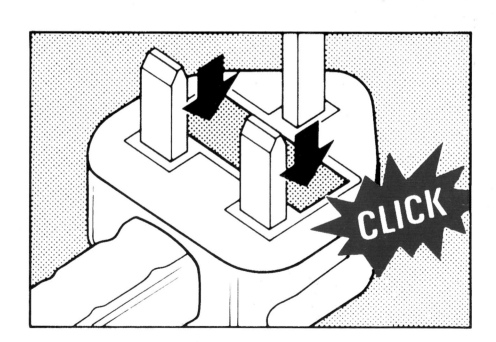

CLICK

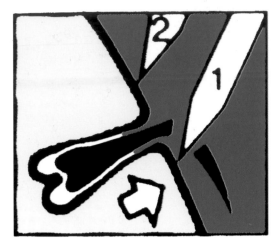
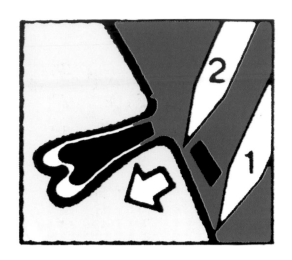

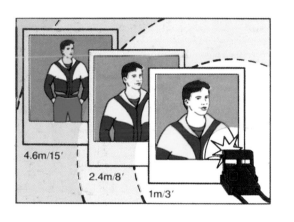

4.6m/15'

2.4m/8'

1m/3'

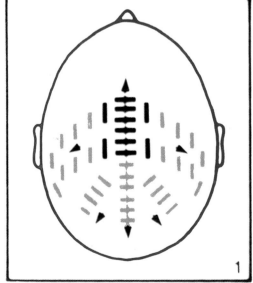
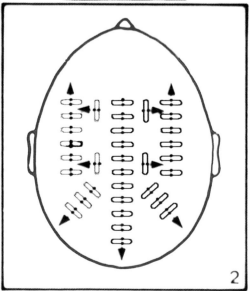

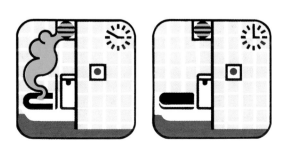

Cranial surgery or hair curler diagrams? Sometimes you have to suffer for the desired result.

The Art of Instructional Design

Pull the string and the fire cracks.

Artist Boudewijn Bjelke pokes fun at visual instructions in this drawing: "The problem. You need. The result."

The Art of Instructional Design

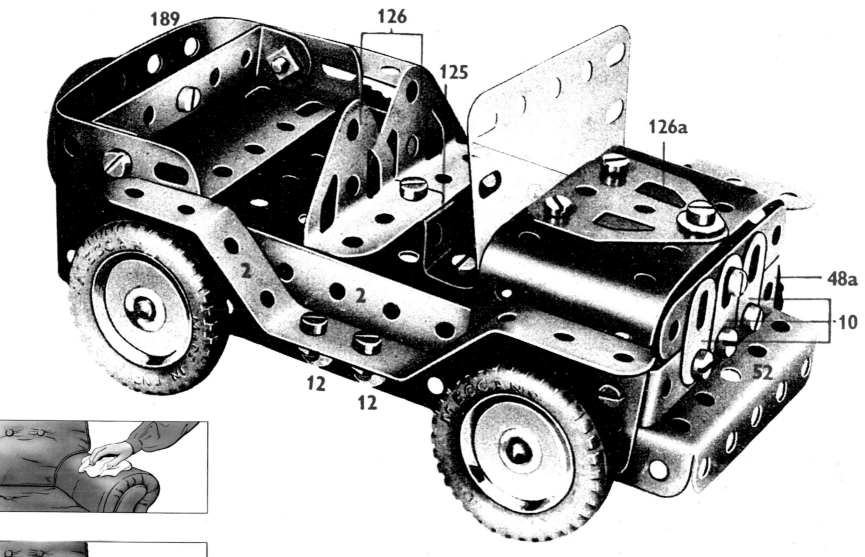

That's What It Should Look Like

Sensation is proportional to the logarithm of the stimulus.

Weber-Fechner's Law, the oldest law in psychology

The cupboard as it should look. A television with a crystal-clear picture on the screen. Shiny shoes. *Nouns again.* The nouns of success. The exclamation marks! You've done it! We're just as surprised as you are.

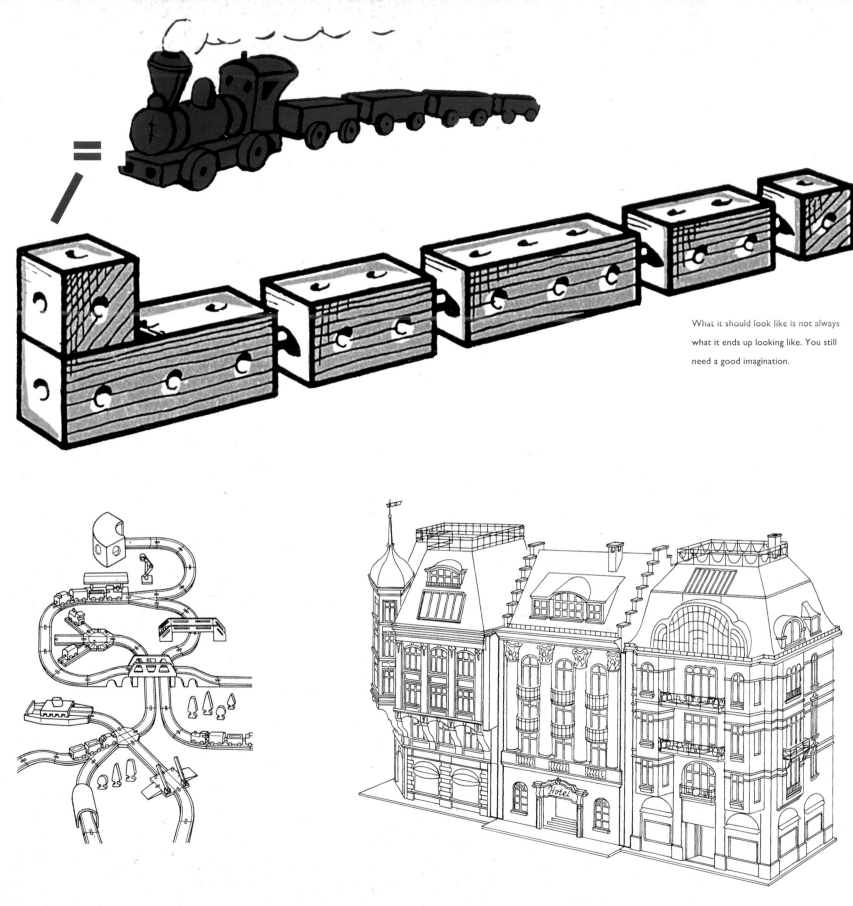

What it should look like is not always what it ends up looking like. You still need a good imagination.

The Art of Instructional Design

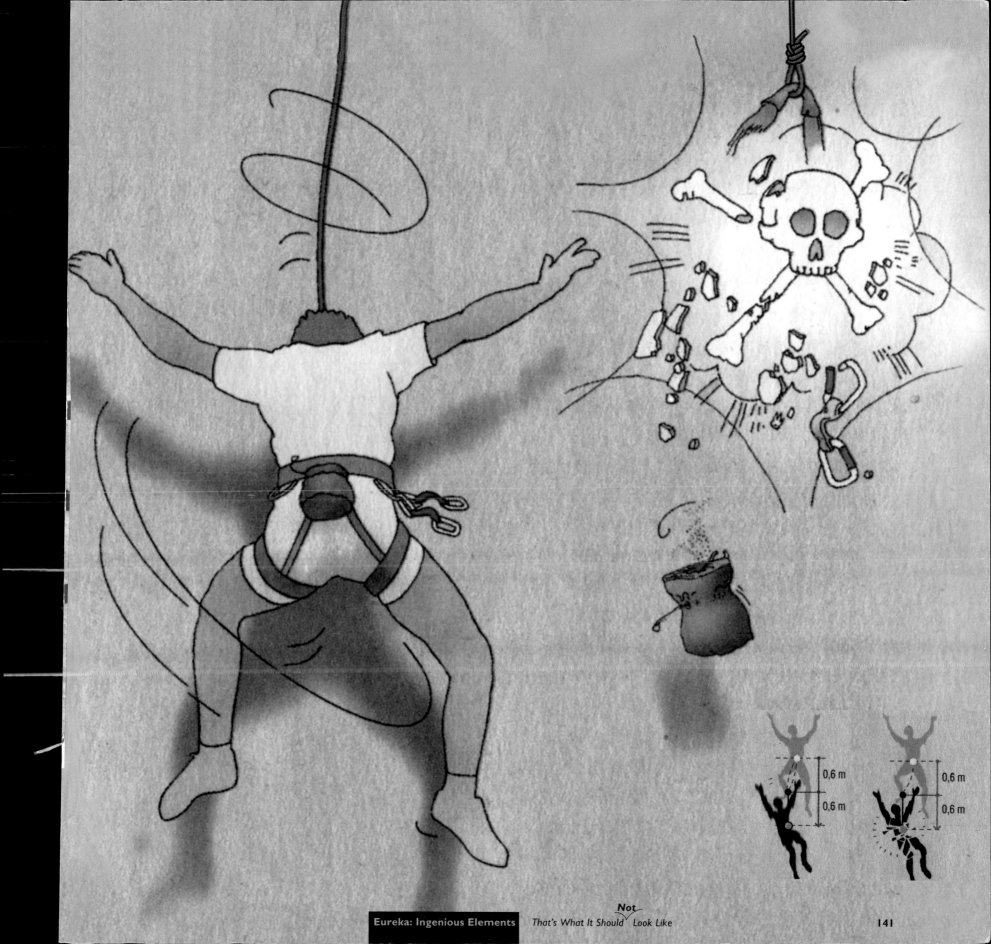

About this Book

In 1981 Paul Mijksenaar designed a calendar featuring items from his collection of user instructions. An exhibition on the theme of instructional design was then developed by Mijksenaar and Piet Westendorp, who also wrote the accompanying book (in Dutch, with a German translation). This exhibition traveled to various cities in Europe and the United States. After its showing at the Fashion Institute of Technology in New York, Andreas Landshoff and Joost Elffers decided to publish a completely new book on the subject of visual instructions: *Open Here*.

Westendorp created a new concept for *Open Here*, added information on the history of user instructions and wrote a new text. Westendorp and Mijksenaar made a fresh selection of illustrations and Mijksenaar designed the book.

The expanded concept and great images make *Open Here* a methodical guideline and splendid source for the designers of visual instructions, as well as a fun read for anyone who's ever had to consult directions for use.

In this book we have left out much everyday art, but also fine art. The work of Tammy Stubbs, for instance. Stubbs is an instructionalist artist who takes Pop to new levels of meaning by questioning the glib forms of behavior modification in our present-day struggle for life.

The Art of Instructional Design

What We Left Out

I doubt whether all mechanical inventions yet made have lightened the day's toil of any human being.

John Stuart Mill

Visual instructions are all around us, not only in manuals, on packaging and on products themselves. In this book we've only discussed the visual instructions for some specific types of products and services. We showed the instructions for one type of Kodak disposable camera; we did not show instructions for using cameras in general, or instructions for taking beautiful photos. Marvelous and ingenious visual instructions can also be found in school books, in technical training books, in guide books for playing chess, using a compass, how to repair your bike and tie knots. In scientific illustrations and professional instructions (for doctors, pilots, gourmet chefs). Even diagrams charting statistical information can be considered as small pieces of great art. We left out all the beautiful on-screen instructional art, the videotapes, the interactive tutorials with "funny" animation, the on-line help. We didn't include timetables or product graphics. Neither did we examine maps, although there are enough gorgeous examples to make a book twice as thick as this one. Even without these, this book should have been twice as long. . . .

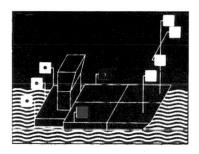

From a handbook on sailing.

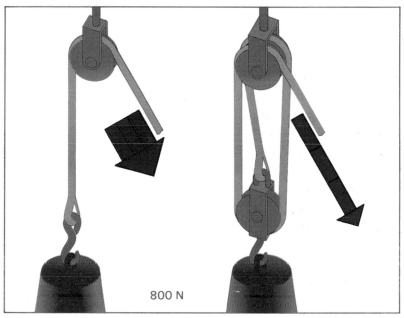

From a book that explains the working of pulleys.

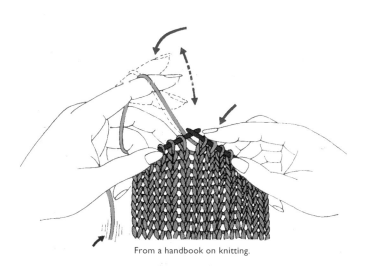

From a handbook on knitting.

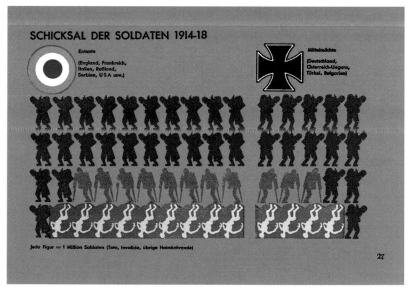

"Infographics": the visual display of statistical information.

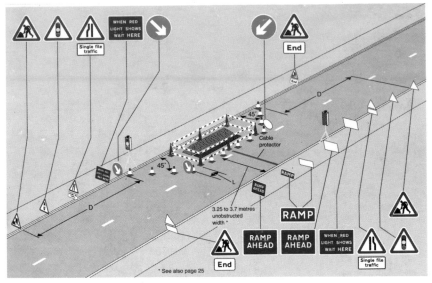

From a manual for road builders: how to indicate a ramp.

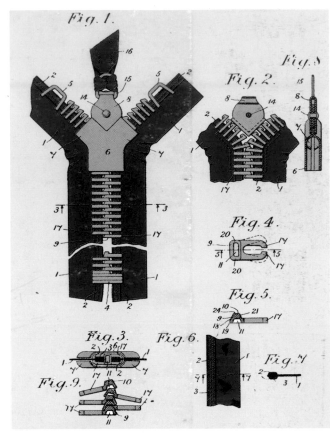

Patent application for the zip fastener, L. Judson, 1893. (Color added.)

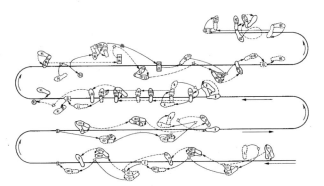

From a T'ai Chi poster.

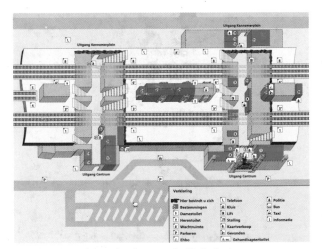

Map of the Amersfoort railroad station, the Netherlands.

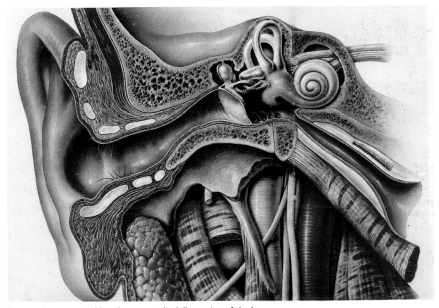

From a pharmaceutical brochure: a medical illustration of the human ear.

The Art of Instructional Design